PEREGRINE BOOKS

TIME, SPACE AND THINGS

'This book is an attempt to survey, in simple terms, what physics has to say about the fundamental structure of the universe. It aims to extract, from a whole range of specialized activities, the basic essential concepts and to present them in plain, non-mathematical language. There are some splendidly bizarre ideas in physics, and it seems a pity to keep them locked up in narrow boxes, available only to a small esoteric crowd of key-holders.'

In this Peregrine Original, Dr Ridley deals with all the basic concepts which physicists use to understand the universe – electro-magnetism, gravity, mass, energy, time, motion, space – and explains the techniques with which they manipulate those concepts. His exposition throughout is elegant and lucid, as he says, 'The mystery of things, the mystery of relations and inter-relations and the mystery of complexity claim our wonder and our involvement.'

B.K. Ridley is Reader in Physics at the University of Essex. He has also been a Visiting Professor at Cornell, Princeton, Stanford and Danish Technical Universities. His previous publications include research papers on the physics of semiconductors.

B. K. RIDLEY

TIME, SPACE AND THINGS

PENGUIN BOOKS

Penguin Books Ltd, Harmondsworth, Middlesex, England
Penguin Books, 625 Madison Avenue, New York, New York 10022, U.S.A.
Penguin Books Australia Ltd, Ringwood, Victoria, Australia
Penguin Books Canada Ltd, 41 Steelcase Road West, Markham, Ontario, Canada
Penguin Books (N.Z.) Ltd, 182–190 Wairau Road, Auckland 10, New Zealand

—

First published 1976
Copyright © B. K. Ridley, 1976

Made and printed in Great Britain by
Hazell Watson & Viney Ltd, Aylesbury, Bucks
Set in Monotype Times

TO SYLVIA

CONTENTS

LIST OF FIGURES

LIST OF TABLES

PREFACE

THIS book is an attempt to survey, in simple terms, what physics has to say about the fundamental structure of the universe. It aims to extract, from a whole range of specialized activities, the basic essential concepts and to present them in plain, non-mathematical language. There are some splendidly bizarre ideas in physics, and it seems a pity to keep them locked up in narrow boxes, available only to a small, esoteric crowd of key-holders. Naturally, in opening the boxes and trying to air the contents without the elegant restraint of mathematics, or without the adumbration of historical context, one runs the risk of pleasing nobody, and laying oneself open to the charge of superficiality on the one hand and incomprehensibility on the other. Nevertheless, the risk is worth taking, qualm-making though it be, if only because there is a super-abundance of specialization, and very little of generalization.

Even within the speciality of physics itself, the student is confronted with so many finely delineated trees that all too often he fails even to appreciate the existence of the wood. Occupied as he is with hacking his way through the thickets of thermodynamics, electromagnetism, quantum mechanics and the rest, he may be forgiven for losing all sense of direction and wishing he could rise above the forest and get his bearings. It was to go some way towards meeting this wish that a short course of lectures was introduced into the undergraduate course in physics at the University of Essex, out of which this book developed. Many concepts, familiarity with which is usually taken for granted at undergraduate level, have been introduced more or less from scratch. Concepts normally encountered only at postgraduate level have been considered to be intrinsically no more difficult than those we come across at school and have been included without ceremony. If they are difficult to apprehend, the difficulty is largely one of

9

plain unfamiliarity. If they were left out, this book would not achieve the aerial view of physics it grasps for.

I feel with Mark Twain:

There is something fascinating about science. One gets such wholesale returns of conjecture out of such a trifling investment of fact.

I would like to think that some of that fascination is transmittable to the general reader, who wants to catch up on his physical universe; to students of the humanities and social sciences, who wish to be generalists; and to sixth formers, who are contemplating specializing in physics. I would also like to think that the book may prove useful to teachers of physics at all levels, and to university students reading physics, as part of their background reading. Having said that, I am all too aware of the difficulties in writing a book which has something to say to both generalist and specialist. I am equally aware that there are far too few attempts to do so, among professional scientists. The cultural gulfs between the intellectual disciplines have never been wider. It is impossible to over-specialize; but never to generalize, is not only possible, it is usual. Those cultural gulfs are unaesthetic, and, at times, downright dangerous; and, if this book goes any part of the way towards bridging them, it will have achieved at least one of its objects.

B. K. RIDLEY

Colchester, July 1974

CHAPTER 1
THINGS

The poet's eye, in a fine frenzy rolling,
Doth glance from heaven to earth, from earth to heaven;
And, as imagination bodies forth
The forms of things unknown, the poet's pen
Turns them to shapes, and gives to airy nothing
A local habitation and a name.

Shakespeare: *A Midsummer Night's Dream*

PHYSICS is about the simple things of the universe. It leaves the complication of life and living objects to biology, and is only too happy to yield to chemistry the exploration of the myriad ways atoms interact with one another. The living cell is clearly an impossibly complex system, and so, for example, is a surface – any surface. There may be the occasional flirtations with these topics, the one in biophysics, the other in chemical physics, but by and large they are terribly difficult to deal with. Cells and surfaces are not simple things.

It could be argued that simple things plainly do not exist. The Queen might boast to Alice that she could think of as many as six impossible things before breakfast, but she would be hard put to it to think of six *simple* things. But, to take a primitive example, what could be simpler than a chunk of rock, a stone that can be picked up in one hand and thrown? There is something very real and immediate about a thing that can be seen and felt and manipulated. And it is vital to have some understanding of how it can be moved around – a stone is the simplest missile, after all. A stone-age war department might have been keen to commission some customer-oriented research on stone ballistics, but few physicists, unless heavily bribed, would touch it. A stone is too complicated. Its shape is horribly irregular, and think how intricate the flow of air would be over its rough surface. Far too messy to sort out what is general for all missiles from what is particular for this stone.

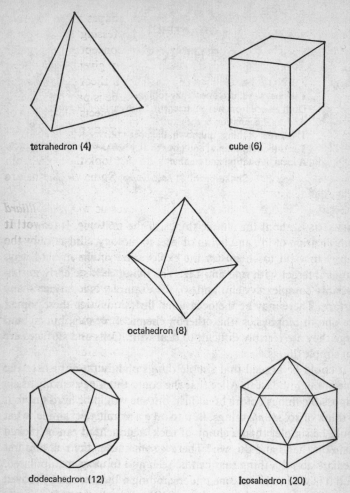

tetrahedron (4)

cube (6)

octahedron (8)

dodecahedron (12)

icosahedron (20)

Figure 1. The five regular solids.

Then clearly one cuts the stone into a regular shape – say one of the five regular solids (Figure 1). An object in the form of a tetrahedron, or a cube, looks a lot simpler to deal with because of the high symmetry. And how simple to have only five regular shapes to think about! Surely objects cut into such shapes must have an especially significant place in a subject professing to deal with simple things. The fact is they have not. The concept of the regular solids is a lovely one and important to apply given the slightest chance. Kepler tried very hard to construct a theory of the solar system on this basis. The symmetry of the cube is particularly important in crystallography. Nevertheless, real objects in the form of one or another of the regular solids do nqt play any role in physics. The reason is that the regular solids suffer from corners and edges. Their regularity is only relative. They do not look the same from whatever point of view one cares to adopt. Some directions are 'more equal' than others.

Getting rid of the corners and edges leaves us with *the billiard ball*. To a physicist a billiard ball is a lovesome thing, God wot! It has an archetypal significance in the subject, unrivalled even by the weightless string. It looks the same from all directions and it can be handled, thrown, swung or rolled to investigate all the laws of mechanics. Our stone-age natural philosopher would have insisted on a grant especially ear-marked for the production of spherical stones. But, in spite of the undoubted glamour of our billiard ball, it is still not simple enough. Its flaw is that it has a surface and, as we mentioned at the beginning, a surface is not a simple thing. Yet any object, if it is to be distinguished from its surroundings, must necessarily possess a surface. That being so, we idealize the surface away by pure imagination – infinitely sharp, perfectly smooth, absolutely featureless. And while we are idealizing, let us make the billiard ball absolutely uniform, of infinitely hard and perfectly elastic material – shall we call it *utopium*? Now there is the first simple thing of physics – a billiard ball made of utopium.

Yet nothing like it exists. The utopium ball is a product of, literally, an ideology, and nobody, being dispassionate, believes entirely in the products of an ideology. That is, paradoxically, its strength. We know from the outset it is wrong, in the strict sense that it cannot possibly be exactly true, and so an assessment of how wrong it is in the particular case can begin straight away. Its conceptual simplicity is invaluable. At one end of the scale the utopium ball can double for a star swinging around a stellar cluster somewhere in a galaxy, or a planet orbiting a sun. At the other end it can be an atom – one of many, perhaps, arranged in the regular lattice of a crystal, or wandering about as the tiniest component of a liquid or gas. Given a positive electric charge it can attempt to pass as a proton; given a negative charge, an electron; or given no charge at all, a neutron. The model of the atom described by Bohr consisted basically of charged utopium particles. Many properties of solids, liquids, gases and plasmas can be quite happily treated in terms of them. In short, the utopium billiard ball is the ideal elementary particle of classical mechanics. Though it is a failure as a useful concept to use in the quantum realm – electrons, protons and neutrons do not behave like billiard balls – its usefulness elsewhere makes it one of the archetypal models in physics.

But nevertheless it is a concept – a product of the imagination. It does not exist out there in the real world clamouring for attention. It is an idea. One day in the future perhaps there will be a subject called Erewhon Physics (to borrow from Samuel Butler), dealing entirely with possible but unreal universes, a subject which has come into being because the physics of the real universe has all been worked out and people still want to carry out the activity. A subject like that would consist purely of things of the imagination and some sort of self-consistent collection of rules. It would really be a branch of mathematics. Real physics deals with things which exist out there in the real world quite independently of our imagination. Yet it uses objects like our utopium ball which look as if they belong to an Erewhonian sort of physics rather than the real world. The reason is that such objects are more than ideas – they are ideals, chosen for their simplicity and used as model starting-points in the process of understanding the character and behaviour of real things. Physics is above all a model-making activity.

While we are on the subject of conceptual things, let us attempt a fine distinction between what we may call physical conceptual things and mathematical conceptual things. The utopium billiard ball is a physical concept. It can be obtained from the real billiard ball by extrapolating a few real properties to the ideal limit. The ideal behaviour differs from the real in degree only, but not in kind. Compare that with another archetypal thing of physics, *the point-particle*, much beloved by theoreticians. The point-particle is a mathematical concept because it is different in kind from a real particle – it has zero extension. It is therefore more unreal than our billiard ball, and we have to be careful not to push the concept too far. Theories of elementary particles are bedevilled by infinities which arise from the enthusiastic application of the point-particle concept. Nevertheless, whenever the extension or the internal structure of a particle is not an important factor, the simple concept of the point-particle is an invaluable tool.

But why concern ourselves with conceptual things when there are so many real identifiable things to be investigated and their natures understood? Unfortunately one cannot avoid a large degree of abstraction. There are just too many separate, unique, real objects in the world to appreciate them as individuals. All we can hope to do is classify into groups and study behaviour which we believe to be common to all members of the groups, and this means abstracting the general from the particular. All sciences function this way, but how successful such abstracting is depends a lot on what is studied. The budding science of sociology has the problem of coping with the effect of individual human beings. Although an amoeba with a strong personality has yet to be discovered, biology has always to keep a wary eye open for the effect of individual living things. Only in the study of inanimate matter can one be really successful at coping with the particular. But in doing so one has to invent ideals like the concepts we have been discussing, trusting that the psychology of the electron as a serious study is a long way off.

Even so, the forms taken by inanimate matter are fantastically varied. How to cope with such complexity?

One attempt is to see all the matter within our immediate experience as just one huge body, the earth. The *Earth* is then our first real thing of physics. The structure and composition of this vast object is incredibly complex but it can behave, nevertheless, in simple ways, almost like a large version of our utopium ball. The *Moon*, the *Sun*, and the individual *planets* also follow as identifiable objects in the same realm. Though each is unique, demanding a field of study all to itself, each is also an example of a general thing. This emerges with the concept of the *Solar System*. The earth becomes one of the planets, and the sun becomes the local *star*, one of 10^{11} stars inhabiting our *Galaxy*, which itself is part of a cluster of galaxies, one of many inhabiting the *Universe*.

Here are all the large-scale real things of physics – the universe, containing all matter, and the only truly unique thing to be studied in physics; the star, atom of the galaxy; the planet, satellite of the star. Universe, galaxy, star and planet all lend themselves to the modelling powers of the chi medium (chi medium is explained later), the endlessly versatile utopium ball, and sometimes even to the blandishments of the point-particle. They represent matter on the grandest scale and, when the scale is so large, one can often afford to overlook the details; but not all the time. In fact the study of stars positively demands a detailed understanding of the structure of matter. So we are directed back to the study of the forms which inanimate matter can take.

In this we are vastly helped by the discovery in chemistry of the *atom*. The simple thing of matter is the atom. And there are only about a hundred different sorts. All the intricate varieties of substance which are presented to us are collections of atoms, of one type, or mixed, or chemically combined with others. The atom is too small to see directly (unless one regards the images displayed by an instrument like the ion microscope as direct viewing) but their existence, supported by overwhelming evidence in chemistry, and in the last hundred years in physics, is not in doubt.

Many of the properties of the matter surrounding us, particularly the mechanical and thermal properties, are understandable in terms of hosts of microscopic billiard balls posing as atoms. Sparsely scattered, they represent a *gas*. Densely packed, with some short-range ordering, they look like a *liquid* or an *amorphous solid* like glass. Impose a long-range ordering and we get *crystals*, liquid crystals as well as solid ones. Put them in chains like a necklace and they become *polymers*. They can hunt around singly, as elements, or in packs, as molecules.

Nature has provided an immense simplification in allowing us

the atom, but it is not enough for physics. Physics craves simplicity, and coping with the detailed properties of each individual
atom, from the simplest – hydrogen – to the most complex, presents itself as a mammoth task, thankfully left to the fortitude of
chemists. Of course, physics cannot escape from this entirely, nor
would it wish to. It must appreciate certain characteristics peculiar
to each atom, but the fewer idiosyncrasies the better. Thus the
physicist is delighted to find that all gases, whatever atoms or
molecules they consist of, behave in a roughly similar way; so
much so that he invents an *ideal gas* as a paradigm. He finds that
fluids tend to flow in certain ways which are quite independent of
their chemical (atomic) properties. He invents the concept of the
perfect crystal, a vision of rows upon rows of atoms, in exemplary
order, yielding properties which depend as much on the symmetry of the arrangement of atoms as on the chemical nature of
them. Rather than be grateful that the number of different sorts of
atom is only a hundred or so, the physicist makes it known he
would have preferred just one – utopium. One is quite enough to
give idealized body to the three states of atomic matter – gases,
liquids, solids.

More accurately, the idealized atom is a beginning of the attempt
to understand the behaviour of matter. However attractive it may
be to evolve general pictures with simple concepts like utopium
billiard balls, one must eventually appreciate the peculiarities of
individual atoms in a real system. How else can one understand
why oxygen melts at $-219°C$ and ice at $0°C$? Although one may
point to the enormous importance of the arrangement rather than
the chemical nature of atoms in a crystal with regard to its properties, and quote with glee the case of carbon atoms which form
the hardest substance known when ordered in a diamond lattice,
and one of the softest, as its use in pencils testifies, when ordered
in a graphite lattice (Figure 2), it is obviously essential that
individual atomic characteristics be ultimately built into whatever
model is developed. A theory of some crystal may begin with a
point-lattice, to explore the role of its special symmetry, and the
points then expanded into utopium balls in order to investigate its
mechanical and thermal properties, but, for increasingly sophisti-

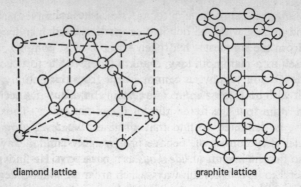

diamond lattice   graphite lattice

Figure 2. Carbon atoms in diamond and graphite.

cated understanding balls have to become real atoms and individual atomic eccentricities have to be taken into account.

That may be so, but it does not stop the physicist, in his quest for simplicity, applying successfully even more unrealistic concepts than the utopium ball or the point-particle. What could be more appalling than to ignore the existence of atoms entirely? Regard this solid or that liquid as a *continuum*, a substance filling all the space occupied by the object. Such a model, in spite of its apparent crudity, works extremely well on a scale much larger than an atom.

The concept of the continuum is, in a sense, the complement of the point. It is just as unphysical, but within its own scope extremely valuable. One of the famous conceptual things of physics is the continuum which is both homogeneous and isotropic. A homogeneous substance has the same properties throughout its extent, independent of position. An isotropic material has the additional simplicity of possessing properties which are totally independent of direction – a material without a grain. Such a medium is so popular in physics that it really ought to have a name – perhaps *chi*, formed from the initial letters (of 'continuum', 'homogeneous' and 'isotropic'), with the Greek letter χ its symbol. Chi appears everywhere – as a model of the Universe itself, as the stuff we have called utopium in the archetypal billiard ball. It makes a splendid gas, a perfect fluid, and, given basic elastic properties, a beautifully simple solid. And as a tremendous mathematical bonus, one can use the differential and integral calculus to describe its behaviour.

But the important advantage of the concept of the continuum over the atom picture is that it emphasizes properties which are to do with the whole of whatever is under scrutiny. Bulk, rather than atomic properties, the holistic, rather than the microscopic, are picked out and given prominence. The detail is deliberately obscured so that the manifestations of the whole stand out clearly. One such manifestation of vital significance in physics is the *wave*.

First appreciated as an up-and-down motion travelling along the surface of water, capable of being reflected, refracted, diffracted and suffering interference, the wave is undoubtedly a thing. In a continuum it arises out of the elastic interaction of one infinitesimal element with its neighbour in a jelly-like way. In an atomic model it arises out of the interactions between atoms, and is therefore a form of collective motion of all the atoms in the material. The important point is that waves, though nothing more than motions of atoms, are as much entities inhabiting matter as atoms themselves are. They are called *elastic*, or *mechanical waves*. They may be *surface waves*, or *bulk waves*. In either, their essential feature is the to-and-fro displacement of matter. Three types can be picked out. One sort is the *longitudinal wave*, familiar to us as sound wave, in which the to-and-fro displacement is along the direction of propagation (Figure 3). Then there are two *transverse waves*, in which the displacement is at right-angles to the direction of propagation, either up-and-down or side-to-side. The longitudinal, or compression-rarefaction, wave can travel through anything except a vacuum. The transverse, or shear waves, can travel through solids only.

An excellent example of mechanical waves is to be found in the study of earthquakes. An earthquake produces vibrations in the surrounding rocks which travel round and through the earth, and these so-called seismic waves can be detected in laboratories all over the world. From the study of bulk seismic waves a great deal has been inferred about the internal structure of the earth. Indeed the existence of a shadow zone for shear waves, in which only com-

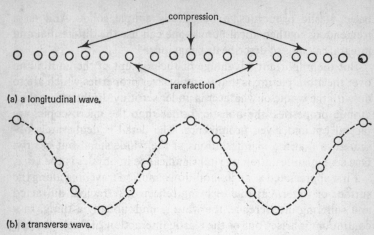

(a) a longitudinal wave.

(b) a transverse wave.

Figure 3. Longitudinally and transversely polarized waves. The dotted line in (b) is a sine curve.

pressional waves can be detected, is remarkable evidence of the earth's liquid core.

If the continuum is complementary to the point, the wave is complementary to the particle. And if the utopium billiard ball is the idealization of a particle, the complementary ideal is the *sinusoidal wave*, which has as much claim to being archetypal in physics as has the billiard ball. But the very description of the ideal wave in terms of a mathematical function seems to call in question any attempts to compare waves and particles. The wave appears to be very different in kind from an atom, for the latter, seen merely as a particle, seems capable of existing without any motion associated with it, whereas the very nature of a wave incorporates motion. Yet one must be careful. More sophisticated models of the atom show it to be full of internal motion. But then it is a fact that the precision with which the frequency and wavelength of a wave may be defined depends upon the spatial and temporal extent of the wave: the greater the extent, the better the wave is defined. Thus, the well-defined wave is spread out over space, while the particle is strictly localized.

20

In practice, this difference is really only one of degree. The important bridging thing here is the *wave-packet*, a travelling vibratory disturbance of limited spatial extent (Figure 4). Indeed, this side of Erewhon physics, we never deal with waves of infinite extent, so, in fact, our concern in practice is always with wave-packets, and these behave in many respects like particles. We have, therefore, the situation that the collective motion of particles can itself behave like a particle, something which is familiar if the motion is purely translational, but is perhaps surprising when the

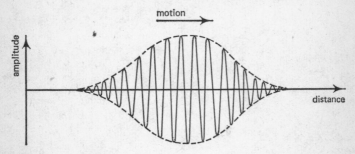

Figure 4. A wave-packet.

motion is vibrational. A tennis ball travelling across the net is an example of the collective translational motion of its atoms, and there is no difficulty in regarding the tennis ball as a particle. To regard the elastic wave generated within the ball as being to some extent like particles needs a little practice.

The idea of the wave-packet is as vital as the idea of the particle, when it comes to understanding the structure of matter. Although each carries its charge of energy and momentum and each occupies a finite amount of space, the behaviour of the wave-packet is very different from that of a classical particle. If two particles come together, they bounce apart. If two packets come together, they merge into each other, causing the amplitude of vibration here to rise and there to disappear. In a word, they interfere. If a particle has to get through a slit, it either does or does not. If a wave has to get through a slit, some of it always does, though its direction of travel afterwards may not be as well-defined as before, because of diffraction. Waves exhibit interference and diffraction, particles do not. Particles are atomic by nature – either one has one or one has not. Waves are jelly-like and plastic, and appear to be capable of being reduced continuously in amplitude to less than the merest discernible train of wobbles.

Nice, simple entities, quite distinguishable in behaviour from one another, mechanical waves and atomic particles are the two splendid things of macroscopic matter, as perceived by classical physics.

CHAPTER 2

STRANGER THINGS

'I can't believe *that*!,' said Alice.

'Can't you?,' the Queen said in a pitying tone. 'Try again: draw a long breath, and shut your eyes.'

Alice laughed. 'There's no use trying,' she said: 'one *can't* believe impossible things.'

'I daresay you haven't had much practice,' said the Queen. 'When I was your age, I always did it for half an hour a day. Why, sometimes I've believed as many as six impossible things before breakfast.'

Lewis Carroll: *Through the Looking Glass*

MATTER has other properties apart from mechanical ones. All matter can be electrified, and some can be magnetized. Instead of solids, liquids and gases, one may classify in terms of conductors, semi-conductors, and insulators, depending on the ease with which an electric current flows. Or one may refer to magnetic behaviour as ferromagnetic, ferrimagnetic, paramagnetic or diamagnetic, depending on a material's response when placed between the poles of a magnet. Or, again, one may focus on electrical polarization and speak of dielectrics and electrets. Other classifications of matter in terms of electromagnetic properties are possible. The variation and richness of electric and magnetic behaviour is immense.

Fortunately, the link between magnetism and electricity is strong. Magnetic effects exist wherever there is electricity in motion. Magnetism is just a manifestation of electric currents. Therefore, to understand electromagnetism we need to concentrate only on electricity and its motion.

Electricity in uniform motion is actually fairly dull. Compass needles in the vicinity of a steady direct current will point away from magnetic north, but that is all. Non-uniform motion is another matter. Alternating currents are the source of one of the most vital entities of the whole of physics, namely the *electromagnetic wave*. Whenever electric charge undergoes acceleration, an electromagnetic wave is produced (Figure 5). Metal aerials carrying electric currents oscillating at modestly low frequencies, such as a few thousand cycles every second, radiate radio waves. At a million times higher frequencies they would radiate microwaves, perhaps as part of a radar system. The only difference between the two sorts of waves is the frequency. Their natures are identical otherwise. Visible light is just another electromagnetic

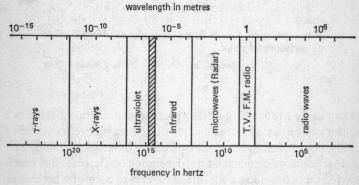

Figure 5. The electromagnetic spectrum. The shaded region is the visible range.

wave. Its frequency is a million times that of radar. In between is the non-visible light, the infrared radiation emitted by everything with a temperature above absolute zero. At higher frequencies are the non-visible ultraviolet rays, X-rays and γ-rays. The essential observation is that a wobbling electric current in one place can induce a wobbling electric current in another place.

All electromagnetic waves travel in vacuum at the same phenomenally high speed ($2 \cdot 997925 \times 10^8$ metres per second), some million times faster than sound travels in air. All have as their source, in some way, the non-uniform motion of electric charge. All can induce a non-uniform motion of electric charge. This, plus their speed, plus their ability to propagate through nothing at all, makes their value in communication paramount. Which means more than transmitting T.V. programmes and talking to astronauts. It means that it is principally through the agency of electromagnetic waves that we learn things about the universe. This is an obvious fact, whose consequences, explored by the theory of relativity, have been nevertheless quite literally explosive.

In many ways electromagnetic waves behave in a homely fashion, similar in kind to mechanical waves. True, the electric currents they induce are always transverse to the direction of propagation, so they behave more like a transverse elastic wave than a longitudinal sound wave. Otherwise all the effects we associate with waves of atoms, which are distinguishing features of waves as distinct from streams of particles, are present in electromagnetic radiation. Thus, electromagnetic radiation passing through a narrow slit will spread out and exhibit all the properties of diffraction that honest-to-God water waves and sound waves exhibit (Figure 6). Superimpose two electromagnetic waves and you will get the phenomenon known as interference – a reinforcement wherever the peaks of the two waves coincide and a cancellation wherever the peak of one wave coincides with the trough of the other. A stream of classical, simple, particles just cannot do that sort of thing. Only waves can. So electromagnetic radiation behaves like a mechanical wave.

But then horrible questions arise. Mechanical waves cannot travel through space devoid of matter. They are a cooperative motion of atoms or molecules, and their velocity is determined by the properties of the matter in which they travel. But electromagnetic radiation travels through a vacuum. So what, materially, is vibrating in empty space? Does the vacuum have a structure? Or are we asking the wrong questions? Clearly there is a new and mysterious thing here.

We will come back to electromagnetic radiation later (not, regrettably, to clear up the mystery but, if anything, to make it worse – electromagnetic radiation is just basically mysterious). First, we must turn to the question of the nature of electricity itself.

As myriad experiments show, electric currents consist of streams of particles. Basically, the particles are of just two kinds, the

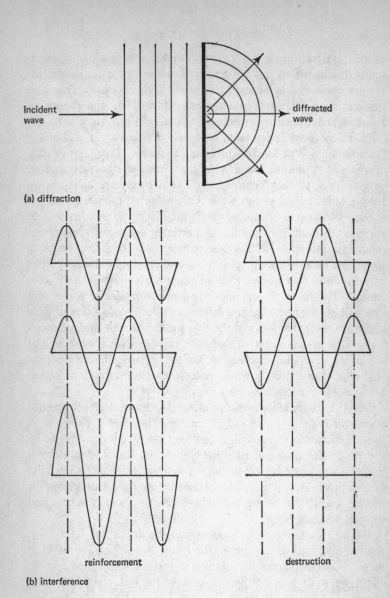

incident wave

diffracted wave

(a) diffraction

reinforcement

destruction

(b) interference

Figure 6. Diffraction and interference.

negatively-charged *electron* and the many-times-heavier, charged atoms, called ions, which are usually positively charged, but can be negatively charged (Figure 7). Of all elements hydrogen is the lightest, and when charged positively it provides the lightest ion known. This is the *proton.* All the electromagnetic phenomena of the world are traceable to the activity of electrons and ions, and ions are just atoms that have lost or gained one or more electrons. Given that the electromagnetic properties of matter almost totally determine the mechanical, this reduction to the electron and ion is an impressive simplicity.

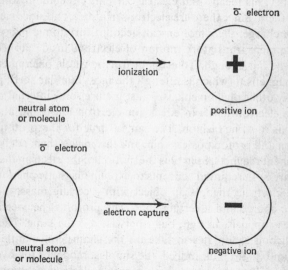

Figure 7. Electrons and ions.

The electric current which flows in our T.V. aerial is a stream of electrons. The billiard-ball image of the atom in the metal disintegrates into two components: a heavy positively-charged metal ion which remains stationary, and one or more easily detachable electrons. The atoms in a semi-conductor lose their electrons less readily, and in an insulator electrons are downright difficult to remove from the atoms. In a fully ionized gas the electron and ion are the two elements of what is referred to as the fourth state of matter, namely, the *plasma*. Imagine heating a solid. It melts and becomes liquid. More heat evaporates it and it becomes a gas. Even more heat causes the atoms of the gas to collide with each other so vigorously that electrons are knocked off, producing positively-charged ions. Electrons plus ions constitute a plasma. One of the most familiar, everyday, plasmas is the sun, which consists principally of hydrogen, or more accurately, a swirling mixture of protons and electrons, the fundamental particles of electricity.

Magnetism tells us something more. Atoms can behave like tiny magnets. This means that electric currents circulate in some way within them. Since there are electrons in atoms, and since electrons carry a charge, their motion can account for atomic magnetism. But the mere translatory motion of electrons inside atoms turns out to be not enough. To account for the whole phenomenon of ferromagnetism (which is after all the most familiar sort), as well as many other phenomena, we must picture some internal motion for the electron itself. In effect, the electron must spin about its own axis in some fashion. The same is true for the proton. Both electron and proton possess spin, and this makes them act like tiny permanent magnets. Spin plus the motion of the electron inside the atom are both required concepts to explain magnetic phenomena.

The electron, then, is an object with a certain mass, a certain electric charge and a certain spin. The proton is heavier and its charge is opposite in sign, but otherwise it is the same. They both look strong candidates to take on the chameleon mantle of the utopium billiard ball. Indeed, the simplest model of the hydrogen atom is a proton nucleus with a single electron orbiting like a planet around the sun. (Unfortunately, when we get down to subatomic scales, we find that our billiard-ball atom is a very poor image of what the atom is really like.) As many experiments show, the atom is mostly empty space. It consists of a tiny nucleus, which is positively charged and possesses most of the mass, and an equal negative charge in the form of electrons surrounding the nucleus and occupying an amount of space which is huge relative to the size of the nucleus. Neils Bohr, by making simple postulates concerning the motion of the electron, found he could produce a highly satisfactory theoretical model for the hydrogen atom. The nucleus of the hydrogen atom is the proton and moving about it in certain prescribed orbits is the electron. The *Bohr atom* is one of the most exquisitely simple things and, even though its success rests upon a set of completely arbitrary postulates, it must always retain a privileged role among the mythological characters which adorn physics.

One of its delightful facets is that it points the way towards building different sorts of atoms out of hydrogen. Thus helium has two protons in its nucleus and two electrons orbiting outside (Figure 8). Lithium has three protons in its nucleus and three orbiting electrons, beryllium four, boron five, carbon six, nitrogen seven, oxygen eight, and so on all the way to uranium with ninety-two and the transuranic elements beyond. All the so-called periodic properties of the elements, discovered by chemists, are explained by the Bohr model. But there is a snag. If the helium atom were just two hydrogen atoms fused together, it should be twice as massive as a single hydrogen atom. In fact it is four times as massive. The discrepancy gets worse as we go to heavier atoms. The weight of uranium relative to hydrogen should be ninety-two, but it is, in fact, 238. There is something else. That something else

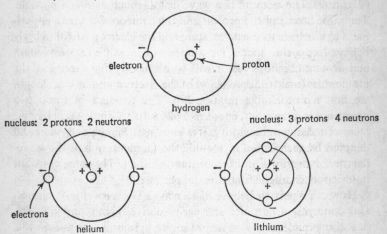

Figure 8. Atomic structure of the lightest elements.

is another particle, the *neutron*. It has very nearly the same mass as the proton, but it is uncharged. Oddly enough, however, it acts like a tiny magnet, just like a proton. It is always present in the nuclei of atoms, hydrogen excepted. Thus helium has two protons and two neutrons in its nucleus and uranium has ninety-two protons and 146 neutrons. Uranium with this number of neutrons is denoted U_{238}, but it is not the only sort. U_{235} exists; that is a uranium nucleus exists with only 143 neutrons in it. It is chemically identical to U_{238}. Chemically identical atoms with different weights are called isotopes, so these things too are encompassed by the Bohr picture with the neutron incorporated. The independent existence of the neutron outside the nucleus has been abundantly verified (as the atomic bomb and nuclear power station will testify). The neutron, the proton and the electron are the things out of which macroscopic matter is made.

What is their nature? Are they tiny billiard balls with spin and charge grafted on to them? Unfortunately for our already over-stimulated imagination, they are not. Besides acting like particles they also behave like waves. Electrons can be diffracted by a narrow slit just as light is. The pattern of electron intensity through a double slit could not possibly be produced by a tiny billiard ball (Figure 9). The electron is a wave just as much as it is a particle. The same is true of the proton and the neutron. Set up an experiment to illustrate its particle nature and the electron will obligingly behave like a tiny speck. Do another to measure its wavelength and an unambiguous result will be obtained. If we measure the momentum (mass times velocity) of the electron and its wavelength we find a remarkable relationship. The product of these two quantities is a constant, equal to $6 \cdot 626 \times 10^{-34}$ Joule-sec. Here we have a fundamental quantity. It is known as *Planck's constant* and denoted by the symbol h. Double the momentum and the wavelength halves to maintain a constant product. The same constant holds good for all the fundamental particles.

How can a particle behave like a wave? The answer is, it cannot. Our concepts of particles and waves derive from chunky, everyday, things and there is no way of seeing in familiar terms how this odd behaviour can come about. The nearest image we can muster

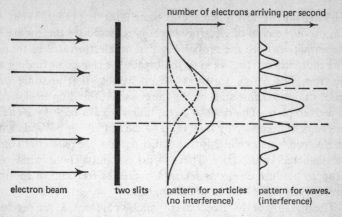

Figure 9. Wavelike behaviour of electrons. The pattern of electron intensity on a plane placed behind the slits would consist of the superposition of single-slit intensities (dotted curves) if the electrons were particles. In fact, the interference pattern is observed, characteristic of waves.

is a wave packet, which has a limited extent in space just like a particle, but which will exhibit diffraction and interference in accord with its wavelike nature. But we can arrange to split a wave-packet in two, perhaps with a fast shutter that drops after half of it has gone through and reflects the other half. Electrons, however, refuse to be split in two – either the electron would get through or it would not. It is a solid entity just like any classical particle, and yet it can give a diffraction pattern exactly like a wave, whose wavelength is Planck's constant divided by its momentum. The electron is neither a particle nor a wave, as we know these terms. It is something altogether different, neither a tiny billiard ball nor a wave-packet.

But since the electron cannot be split, and possesses measure-able attributes like mass and spin, it is difficult to see it as anything else but a particle. How then do we interpret the wavelike patterns? One way is to regard them as being produced by waves which define the possible paths that an electron can follow. In the case of real waves the square of the amplitude at a point determines the intensity, or how much energy there is, at that point. What we can do in the case of electron waves is to associate the square of the amplitude with the probability that an electron will be found. The more intense the wave is, the bigger the chance of finding an electron. The waves are waves of probability that in some way reach out and define statistically the chances of finding an electron in a given place. The path of a real billiard ball struck by a cue is determined precisely by how it was struck. Playing billiards with an electron would be highly traumatic, because the path could only be predicted statistically. There is no certainty about what will happen, because electrons retain a degree of freedom of motion which seems to be built in to their nature.

The measure of this freedom is Planck's constant, h, the product of momentum and a distance which we interpret as the wavelength. In classical mechanics the product of momentum and distance is a quantity known as *action*. Electrons and other fundamental particles appear to possess an intrinsic amount of action, h. Here is a very strange thing, perhaps the strangest of all in physics. Its name is the *quantum of action*, the measure of the intrinsic freedom of motion of the fundamental particles. It is very tiny, so we do not appreciate its existence in objects of ordinary size, but it is there, markedly determining the behaviour of all the particles which make up ordinary-sized objects.

It is even there in electromagnetic radiation. In the photoelectric effect, light of sufficiently high frequency knocks electrons out of solids. Enough energy is absorbed by the electrons from the light to enable them to escape from the atoms near the surface. But if the frequency of the light is too low no electrons are liberated even when we use intense light. The total amount of energy in the light is irrelevant. The effect is determined by the frequency. This effect, and many other phenomena, are explained only if the energy in electromagnetic radiation comes in packets. Electro-magnetic radiation consists of streams of particles! Here we have again a wave–particle paradox.

If we measure the energy in each packet, and also determine the period of the wave (the inverse of its frequency), then we find that the product is, astonishingly, Planck's constant. The product of energy and time is, like the product of momentum and distance, an action. Particles of electromagnetic radiation possess the same quantum of action as electrons, protons and neutrons. Their wave nature is the same as that of material particles – waves of probability. We call them *photons*. In radio waves the energy of a photon is very tiny and hardly noticeable, but in γ-rays the photon energy is large and the particle-like nature stands out markedly in any experiment. Incidentally, the existence of the photon makes it easier for us to think about light travelling through a vacuum. After all, we are happy about electrons streaming through the vacuum in a television tube. Photons join the list of quantum particles along with the electron, the proton and the neutron, particles possessing a freedom of motion constrained by the quantum of action.

Even more bizarrely, the quantum of action stretches out to encompass all the waves that arise out of the collective motion of matter. Sound waves, and other elastic waves, are streams of quantum particles called *phonons*. Even the energy of a straightforward to-and-fro oscillation comes in packets called *quanta*. These entities all share with the elementary particles the statistical element, the uncertainty of prediction, but their detailed properties derive from the properties of the particular bit of matter to which they belong. This distinguishes them from the elementary particles whose properties cannot be derived from anything as far as we can see (which is why we call them fundamental particles).

One casualty of the world of probability waves is the Bohr model of the atom. In the new model the nucleus remains for most purposes a billiard ball, but the electronic orbits disappear and are replaced by patterns of standing electron waves. All but one of the arbitrary assumptions adopted are seen to arise out of the existence of h. The one exception is of some importance, since it highlights the gulf which lies between a particle like the electron, and a particle like the photon. It is Pauli's exclusion principle, and it points to a great division of the elementary particles, besides being of vital importance to our understanding of atoms and of macroscopic matter.

Pauli's exclusion principle states that no two electrons can have identical properties in the same region of space. A given standing wave pattern in the hydrogen atom can be occupied by two electrons, only if their spins are pointing in opposite directions. The two electrons are then not identical in properties. Since spins can be either aligned or anti-aligned, there is no way a third electron can occupy that pattern without violating the exclusion principle. Protons and neutrons also obey the principle. In large numbers they are described by a scheme of statistics developed by Fermi and Dirac. Such particles are called *fermions*.

On the other hand we have particles, like the photon, which behave quite differently. Any number of photons can have identical properties in the same region. The presence of one, far from excluding, positively encourages the company of another. In large numbers such particles obey statistics developed by Bose and Einstein. They are called *bosons*. If bosons are the togetherness particles, fermions are the individualists. The most familiar boson is the photon, quantum of the electromagnetic field. Another familiar field is that of gravitation. The quantum of this field (a purely theoretical construct at present) is known as the *graviton* and is also a boson.

Can we in our quest for simple things reduce all the manifestations of the universe to the interplay of just five elementary particles, the electron, the proton, the neutron, the photon, the graviton? It would be nice if we could, but we cannot. For a start, protons in nuclear matter would fly apart, if it were not for an entirely different force of attraction which exists between them at short range – the so-called strong interaction. The quantum particle associated with this field is the *π-meson*, or pion, a particle about 300 times heavier than the electron. Positively charged, negatively charged and neutral pions have been observed. Heavier mesons, viz: the *K-meson* (Kaon) and the *η-meson* (eta-meson), also exist. These mesons are all bosons. On its own a charged pion exists for less than 10^{-7} second before decaying into a *muon* (μ-meson, 207 times heavier than the electron) plus a neutral massless particle called the *neutrino*. The decay of the muon into an electron also produces a neutrino. This remarkable particle is something with spin which travels at the speed of light, though it is not a photon.

34

At the other end of the mass scale several particles heavier than the proton and neutron are produced by fast proton beams hitting a target. These are the Λ (lambda), Σ (sigma), Ξ (xi), and Ω (omega) particles, known collectively as *hyperons*. Although they exist for only about 10^{-10} second before decaying spontaneously into less massive particles, this does not make them any the less real. After all, 10^{-10} second is an eternity on a nuclear time scale, which is set by the time taken for light to cross a nucleus, viz: 10^{-22} second.

Add to every particle its *anti-particle* – the *positron* as anti-particle to the electron, the anti-neutron and so on. Add to this the extraordinary fact that muon neutrinos are quite different from the type produced in the decay of a neutron into a proton and an electron (β-decay) so that, including anti-particles, there are no less than four distinct sorts of neutrino. Add further to this the transient particles, particles, that is, which live for less than 10^{-15} second, and the total list of sub-microscopic entities looks formidable indeed.

At present all we can do is classify (Table 1). In order of mass we have the four major groups: (i) the *massless bosons* (graviton, photon); (ii) the *leptons* (neutrino, μ (mu)-neutrino, electron, muon); (iii) the *mesons* (π, κ, η); (iv) the *baryons*, subdivided into the *nucleons* (proton, neutron) and the *hyperons* (Λ, Σ, Ξ, Ω). These are the particles stable on the nuclear time scale. If we pick out particles which are entirely stable, as far as we can tell, we would be left with the graviton, the photon, the neutrinos, the electron and the proton. Perhaps these are the elements from which all other particles are made in some way. Some believe that all particles are made up of other particles hitherto undiscovered. The scheme proposed by Gell-Mann invokes three *quarks*: Q_1, with a charge of $+\frac{2}{3}e$ (where e is the elementary charge) and Q_2 and Q_3, each with a charge of $-\frac{1}{3}e$. All have a spin like the electron and there is an antiquark for each. There may even be a second set of quarks known whimsically as charmed quarks. Perhaps quarks really exist and will be discovered one day. It may be that they already have been discovered. A proton appears to scatter fast particles, as if it were made up of three particles stuck together with what theoreticians are calling 'gluon'.

Table 1. The elementary particles

Group		Particle	Symbols for Particles and Anti-particles	Mass (relative to electron)
Long-range field particles		Graviton	(purely theoretical)	0
		Photon	γ	0
Leptons		Electron Neutrino	$\nu_e, \overline{\nu}_e$	0
		Muon Neutrino	$\nu_\mu, \overline{\nu}_\mu$	0
		Electron	e^-, e^+	1
		Muon	μ^-, μ^+	207
Mesons		Pions	π^+, π^0, π^-	~ 270
		Kaon	$\kappa^+, \kappa^-, \kappa^0, \overline{\kappa^0}$	~ 970
		Eta-meson	η^0	1,074
Baryons	Nucleons	Proton	$p^+, \overline{p}^-$	1,836
		Neutron	$n^0, \overline{n}^0$	1,839
	Hyperons	Lambda	$\Lambda^0, \overline{\Lambda^0}$	2,183
		Sigma	$\begin{cases} \Sigma^+, \Sigma^0, \Sigma^- \\ \overline{\Sigma^+}, \overline{\Sigma^0}, \overline{\Sigma^-} \end{cases}$	$\sim 2,340$
		Xi	$\Xi^0, \Xi^-, \overline{\Xi}^0, \overline{\Xi}^+$	$\sim 2,580$
		Omega	Ω^-, Ω^+	3,272

$^+$ denotes positive electric charge

$^-$ denotes negative electric charge

0 denotes neutral particle

Bar over symbol denotes anti-particle. The photon is its own anti-particle, and so is π^0 and η^0.

One cannot deny a sense of unreality when one contemplates such a list of sub-nuclear things. Are not these things merely tracks in a bubble chamber, placed strategically in the vicinity of a vast accelerator (built at enormous expense)? What possible interest can they evoke? Electrons, protons, neutrons, photons, these are all right. They can explain macroscopic matter pretty thoroughly. Why bother with the rest when they are so far removed from everyday things?

The fact is that they *are* everyday things. As you read this, particles of unimaginable energy are pouring into the earth's atmosphere and knocking nuclei to bits. Some of those bits are going right through you or bashing into one of your own private nuclei at this moment. We are continually being irradiated by these so-called *cosmic rays* which consist of, and produce in the air, the elementary particles we created in the laboratory. So they are real things and everyday things. The evolution of life forms depends upon them. We just do not see them very easily.

Nor is their influence confined to the sub-microscopic. They appear as vital elements in the grand scale of stars, not only as the sources of stellar energy and the means by which the chemical elements are produced, but also in vast aggregates as stars themselves. When all nuclear energy is exhausted a star collapses under its own gravitational attraction. Sometimes the pressures become so great that electrons are forced into protons and a *neutron star* is formed. Neutron stars may be the source of the spiky radio emission associated with what the radio-astronomers call *pulsars*. At even higher pressures *hyperon stars* may form.

We have the possibility of elementary particles appearing as vast agglomerations of immense density. Going from the smallest possible scale of things we have jumped to the largest. And here let us identify a final thing, certainly one of the strangest in the whole of physics.

This thing is a body whose behaviour does not depend on whether it is made of neutrons or hyperons or whatever. Its properties stem from the fact that it has mass confined within a critical radius, so concentrated, in fact, that the gravitational field near its surface is huge enough to capture light. Such bodies can neither emit nor reflect light. They are called '*black holes*'. Whatever goes into them can never come out. Some day one may be discovered among the stars, hopefully not *too* close to us.

This account of the things of physics has stretched from the billiard ball and the wave to the neutrino and the black hole. In it we have tried to isolate the simple features of the universe, though not easily understandable ones. The idea has been to point at the important objects much as one would point at items in an exhibition, and say, 'Look, physics is about these things.' And there they are, an example of each, displayed in their glass cases; collectors' pieces of the universe. It sounds a simple enough plan but there have been difficulties. It is worth examining why.

There is, of course, always the problem of selection. It is impossible to be comprehensive, nor would one wish to be. Fortunately, in physics, the fundamental things, being the simplest in some sense, select themselves. More subtly, there is always the danger, inherent in all analyses of wholes into parts, of missing the deep unifying essence by itemizing and dividing into little bits. Mystics will insist upon the underlying unity of all things, however diverse they appear to be. Nevertheless, with our present miserably tiny understanding of the universe, we would do better to agree with Oscar Wilde that 'It is the superficialities that count' and continue to regard the electron and the proton as two distinct entities, at least for the present. The most serious problem, however, has been to separate the things of physics from the rest of physics, and we have quite patently failed.

Even at the macroscopic, everyday, level we have failed to exclude motion from the identification of a thing. The billiard ball is all right. It can inhabit its display case quite happily. But the wave is utterly different. The whole nature of a wave involves motion. The wave is a thing which cannot exist independently of what motion is. And before we can say what motion is we have to talk about what time is and what place is. Things are much worse at the quantum level. We cannot even begin to identify sub-microscopic things without first appreciating that the quantum of action exists. The elementary particles cannot exist independently of what energy and momentum are. Worse, they cannot exist as things independently of their interactions with other particles. Thus the electron has a charge which is a measure of its interaction with photons. The π-meson has a mass which is a measure of the range of the strong nuclear attraction.

In short, the elementary particles cannot be put neatly into separate display cases and labelled confidently as this or that.

Their natures are intimately bound up with their interactions with one another. Interactions involve the dynamic quantities energy and momentum (linear and angular), and these in turn involve mass and space and time. And now we really get into trouble, because to learn about mass and space and time we have to use elementary particles and their interactions with one another. Some of their intricate interdependence is understood, a lot is not. Enough for the present to say that whatever the things of the universe may be, they are certainly not like the utopium billiard ball, isolated from everything else by the glass case in which it sits.

CHAPTER 3

SPACE

A dark
Illimitable ocean without bound,
Without dimension, where length, breadth and highth,
And time and place are lost.

Milton: *Paradise Lost*

THE job of identifying the things of nature is only the beginning. Objects like billiard balls have certain properties which do not have anything to do with the fact that they are billiard balls, properties which they share with chunks of rock and puffs of smoke, properties which stem from the fact that they *are* identities rather than particular sorts of identities. These properties are endurance, size and position. A billiard ball will last for weeks or years until someone smashes it with a hammer, it will occupy a certain volume of space, and will be in this place rather than that. These attributes are very familiar to us. We too endure for a time, have size, and occupy positions. They are clearly basic to the understanding of the fabric of the universe. Indeed they appear in some sense to *be* the fabric on which all the intricate patterns of nature are woven.

But we must be careful. Our familiarity with these things is dangerous, almost as much as our capacity for making them abstract and divorced from reality. Consider the danger of familiarity. It seems clear that an object cannot be in two places at once; but an electron suffering diffraction can. It also seems clear that though size and position is infinitely variable, everything shares the same time; but, as Einstein showed, this is not so. We must check our intuitive ideas all the time. Note that size and position have endurance. If position is without endurance then we are dealing with motion, which is another familiar attribute but a far from easy one to understand. Motion, moreover, is an integral part of the identity of many of the things of physics, as we discovered in the previous chapter, so we need to know about it. Yet, familiar though it is, we must explore carefully the relationship between motion and space and time. It is essential to break loose from the shackles of the familiar and start asking questions like, 'If position can endure is there a sense in which an instant can occupy space?' or 'Is there any sense outside mysticism and metaphor in which a period of time can endure?'

Some questions will be very productive, others will be meaningless, but it is worth risking a few of the meaningless sorts just to get one of the interesting ones. This will, at least, get us out of the realm of chintzy homeliness. But we have to avoid the danger of abstraction. Consider the risk of regarding endurance, size and position as a sort of fabric of the universe. Another way of putting this is that the objects of physics are to be found· embedded in Time and Space – and the capital letters here are no accident. If one does this, one runs the risk of thinking in terms of a Time and Space which has an existence all on its own, independent of the existence of objects.

This is a powerful abstraction not easy to get rid of. Surely Space, or the Vacuum, or whatever one calls it, exists out there surrounding all the things we have identified. Surely Time flows, in some sense, throughout the universe at a steady even rate (whatever that means). It is true that the ancients thought of absolute Space – a hierarchy of position, with man at the centre of the universe. It is also true that the idea of absolute Time persisted right up to this century. But absolute Space and Time really do not exist.

Imagine a billiard ball as the only inhabitant of a universe. What position does it have? The question has no meaning, for position can only be defined with respect to another position, which we call an origin, and there is nothing to define where the origin is. What size is it? But we have nothing to compare it with, so what possible answer could there be? Surely it endures. Since it does not change; since nothing can happen, nothing coming along and bouncing off it, how can we tell the passage of time? We cannot. Both Space and Time have no meaning whatsoever. So much for absolutes. We can get rid of the capital letters along with our prejudices.

For the moment let us concentrate on position and leave time for later. Suppose we have a lot of billiard balls scattered about and, to make things simple, let us suppose each one is stationary. We can now try and define the position of one relative to all the others.

Since none of them is moving, the simplest thing to do is to give each one a number. For example, if we have a system with twenty balls then we can use the integers one to twenty to label positions. Position number three is where ball number three is, position number nineteen is where ball number nineteen is, and so on. Positions in between do not need to be labelled, since our miserable little universe contains twenty stationary billiard balls only. The same procedure could be adopted even with a hundred or a million billiard balls. The important thing is that they do not move.

Of course, once they are allowed to move, that system is useless, because intermediate positions become possible. It then becomes necessary to label any *possible* position that a billiard ball could occupy. A degree of abstraction therefore becomes inevitable.

Suppose all the balls have to roll along one tram-line. We could still use the integers to describe position along the line if we invented some unit. This unit would become our measuring rod. We can imagine identical rods laid along the track, one of them arbitrarily labelled zero, those to the right labelled successively $+1$, $+2$, and so on, and those to the left labelled successively -1, -2, and so on. Any ball on the line has its position defined by an arithmetic sign denoting direction and a number denoting the number of unit rods from zero. If greater accuracy is desired, it may be possible to produce sub-standards of length by defining and physically making, say, a millionth part of the unit rod. The limits to such production of sub-standards is something we will discuss later. (If physical limitation is ignored it is easy to jump to the continuum concept of the tram-line with its infinitesimally close possible positions, which is an idea both useful and dangerous.)

Denoting position by a single number is very simple. Can we apply it to the case of, say, billiard balls on a billiard table? There seems nothing to stop us. Imagine a tram-line winding its way backwards and forwards across the table so that all possible positions are encompassed by the single track. Then the position of a ball is determined by a single number as before. If position is the only thing of interest then this system is perfectly adequate.

But position is not the only factor. In order to describe motion it is also important to understand how a billiard ball gets from one place to another. Another way of putting this is to say that it is important to know how one position is connected to another. We have to describe not only position but also connectivity.

Suppose the billiard table had 100 units of tram-line across, 100 back, 100 across and so on, winding to cover the full length of the table (Figure 10). Position fifty would be adjacent to position 150. If a tram-line world really existed a billiard ball at position fifty would have to traverse all the positions fifty-one, fifty-two, etc. to get to position 150. This is not what is observed. Certainly the ball *can* follow that path. But it can also move directly from fifty to 150. It has physically a further degree of freedom since fifty is connected directly to 150.

To take this into account it is better to think about labelling a position by imagining going there from the zero position. Thus we need two sorts of tram-line X and Y which intersect and provide at every position a choice of two movements. We call the single tram-line a one-dimensional space, and the double set of tram-lines a two-dimensional space. Position 150 becomes position (fifty, one), that is, go fifty units along the X line and 1 unit along the Y line that runs through position (fifty, one). The two numbers describe both position and how a position is connected to adjacent positions.

Now imagine a room filled completely with a continuous sheet folded back on itself innumerable times. Two numbers would be once again sufficient if only position and movement on the sheet is of importance. But we know that a further degree of freedom enters via the connectivity between adjacent folds. Each position holds three choices, so there have to be three sets of imaginary tram-lines and therefore three numbers. Thus our position on the sheet (fifty, one) four folds up becomes (fifty, one, four), a position in three-dimensional space. These three numbers not only specify position but also tell us that a physical body has three degrees of

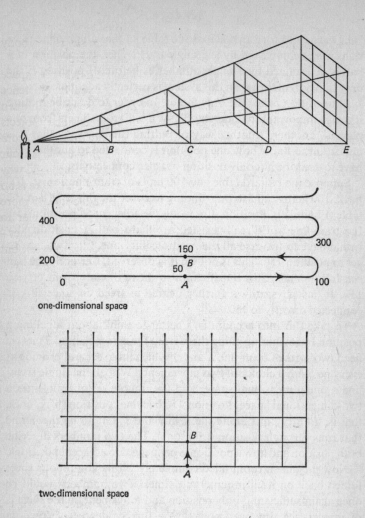

one-dimensional space

two-dimensional space

Figure 10. Connectivity. To go from *A* to *B* in a one-dimensional space we have to follow the track from position 50, through position 100, to position 150. Two-dimensions mean that we can take the short-cut. The uppermost figure depicts the intimate relation between the ubiquitous inverse-square law and the existence of three dimensions.

freedom of movement. Roughly speaking, we can say that a body in any position has a choice of moving along, up or sideways.

Perhaps there is a further choice of movement in a fourth space dimension. If there is we have not discovered it yet. Its existence would produce some rather strange effects because of the hidden connection between familiar positions in three dimensions. Objects moving along the new dimension would suddenly disappear from one place and reappear at another, without having traversed our familiar space at all. The theme is rich for science fiction. And why not a fifth or sixth dimension? The dusty answer is that there is no observational evidence which requires the introduction of more than three space dimensions. Three dimensions are simpler to handle than four. Moreover they are all equivalent to one another – it does not matter how we set up our three sets of tram-lines; but the fourth would be quite different to the other three. We would lose isotropy, the property that one direction is as good as another. So we will stick to our 3-D world until unsolved problems force us to do otherwise.

Isotropy, after all, is precious. It enables us to set up a simple frame of reference to suit the problem. Thus in one case we chose three axes (tram-lines) at right-angles to one another and label them X, Y and Z, and we can let the X axis point in any direction we like. In another case we may find another type of coordinate system more appropriate (Figure 11). Whatever we choose it will not prejudice the basic physics we are dealing with. Our ideal of space is, after all, a 'chi' medium – a continuum which is homogeneous and isotropic. In it there is no preferred place or direction. The only constraint on the coordinate system is that it possesses three independent quantities available for denoting position.

But there is more to do than just denote position. There are measurements of length and area and volume to be made. Straight lines have to be defined, properties of triangles have to be explored – in short, the geometry of our space has to be discovered. In mathematics all sorts of possible geometries can be imagined. In science, which particular geometry fits the real world must be discovered by experiment.

Let us begin by measuring the distance between two billiard

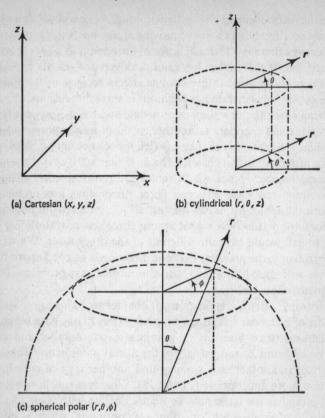

(a) Cartesian (x, y, z)

(b) cylindrical (r, θ, z)

(c) spherical polar (r, θ, φ)

Figure 11. Coordinate systems.

balls A and B. We take our standard rod, choose some way of walking from A to B, and note down how many unit rods it took to cover the path. The next day we do it again, this time going by a place called C (where the view is rather good, perhaps). We discover more rods were needed than before. We repeat the experiment many times, choosing a different path between A and B each time. Many different distances are recorded. Are A and B moving or does the distance apart depend upon the path? After gazing at our results we conclude that the latter is the simpler explanation and, moreover, we discover a special path between A and B for which the number of rods is a minimum. This number is defined as *the* distance between A and B and the path taken is defined as a straight line. Note the lack of appeal to intuition, or the intrinsic nature of things, or whatever. The straight line is to be determined by experiment and a procedure is defined.

We can investigate our geometry further by constructing a circle and measuring the relation between the radius r and circumference l. We discover that $l = 2\pi r$, where 2π is a constant quantity for all circles we have measured up to now. We can divide the circumference into equal units, draw lines to the centre, and define what we mean by an angle. Constructing and measuring triangles we discover that the sum of the three angles is 180°. We discover the laws of trigonometry, which becomes useful when we have to determine distances without being able to walk between A and B. Everything fits with the mathematical geometry invented by Euclid. We speak of our space as being Euclidean. This is so familiar: why bother repeating it? Because familiarity is soporific and it could have turned out quite differently, and may yet do so.

Consider the case of the expanding triangle, which so startled that strange world known to us as Sphereyland. The inhabitants of Sphereyland are called Sphereys. They live in a two-dimensional world occupying the surface of a perfect sphere. In many ways their experience is the same as ours, since we also live on a globe, but unlike us they have no knowledge of the third dimension of up-and-down. This is of mutual advantage, since it allows them to lead a simpler life and it allows us to feel superior. For a long time the limited geographical horizons of the civilized Spherey nations made Euclidean geometry extremely acceptable. Small triangles, small circles, all had the expected properties. 'We live in flat world' they said, and believed it.

One day a Spherey physicist, with nothing to do except brood on the flatness of things, wondered whether triangles made bigger would still be Euclidean. His first request for a grant was turned down, since the country was going through a difficult period as regards finance, and expanding triangles did not look marketable. His second succeeded, since someone thought that expanding triangles might conceivably be a weapon, though no one was sure how. He soon discovered that as triangles were made larger the sum of the angles increased. Turning to circles he discovered that the circumference of a large circle was appreciably less than $2\pi r$. Moreover there existed a maximum size of triangle, and of circle. Beyond a certain radius, increasing the radius resulted in a *smaller* circle. He published these remarkable results and was lynched by the Keep Sphereyland Flat Society. Now, of course, his conclusion that space is not flat but curved, is accepted by everyone (except of course, the KSF Society).

From our superior position we can see how these extraordinary results come about (Figure 12). The Spherey straight line is, to us, a great circle whose centre is at the centre of the sphere. A triangle is formed by the intersections of three great circles. The triangle with its apex at the North Pole and its base stretching a quarter of the way around the equator has each angle equal to 90°. The equator itself is the largest possible circle, its radius the distance to the pole. We can absorb these facts within a framework of three-dimensional Euclidean space. The Sphereys do not have that advantage and therefore must work with curved space, for, to them, space *is* curved.

Another non-Euclidean geometry is that found on a saddle-like surface, on which the straight line looks to us like a hyperbola (Figure 12). The Sphereyland space is closed in the sense that if one travels for a long time in a straight line one arrives back at the start. Saddle space is open, since hyperbolae never join their ends. The sum of angles of a large triangle in this space is *less* than 180°. But as in Sphereyland, as indeed in our own case, on a small scale space is Euclidean to a good approximation. Many physicists think that our own space may be curved on the cosmological scale, and in the vicinity of intense gravitational fields. Only experiment will tell us if this is so.

It is important to know what geometry to use, particularly if trigonometry is involved. Suppose we wish to measure the distance from a point on the plain to a mountain peak. There is no material path along which to lay our measuring rod. The simplest solution is to use light. After all, we see the mountain peak, be-

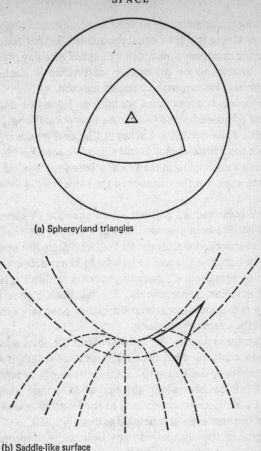

(a) Sphereyland triangles

(b) Saddle-like surface

Figure 12. Non-Euclidean surfaces.

cause light has come from it to our eyes, so why not use light to determine distance. All we have to do is to satisfy ourselves that light travels in straight lines as defined by the least distance approach involving measuring rods. As long as the medium through which light travels is uniform then light does travel in straight lines. We can therefore lay out a base line, measure it in the conventional way, and observe the mountain peak from each end of it, noting the angle made by the line of sight and the base line in each case. Using the laws of trigonometry for flat space, we can compute the distance. This result will agree with any other method we care to use, so on an earthly scale space is Euclidean. The method used is known as the parallax method.

The problem is acute when we wish to know the distance to a star. First of all we have to adopt an operational approach based upon light. Thus we define a straight line as the path traversed by light. (We have little choice about this.) We then use the diameter of the earth's orbit around the sun as a base line, assume that space is flat, and work out the distance in the same way as we did for the peak.

If space is curved we will get a wrong result. A superior being with knowledge of a fourth-dimension would see that our light rays in fact travelled along curves. We, who are without experience in a fourth dimension, have to let a light beam define our straight lines and be prepared, if necessary, to work within the framework of non-Euclidean geometry. So far, we believe that Euclidean geometry is a good approximation except near large masses and except on the cosmological scale.

So much for geometry. Now, what about the unit measuring rod we introduced some time back? The unit of length is called the metre. Once it was defined as 10^{-7} of the distance between the North Pole and the Equator along the line of longitude which passed through Paris. Then it was defined as the length of a certain bar of platinum-iridium. Now it is defined as exactly 1,650,763·73 vacuum wavelengths of the orange-red light emitted by krypton-86. If we are happy about letting light define straight lines we may as well go the whole way and let the wavelength of a good monochromatic beam serve as the standard for length. In fact, we let hundredths of this wavelength be the smallest unit, as the quotation of the number to two places of decimals shows. This means that, if our experimental technique is good enough, we can measure directly any length from one hundredth of a wavelength of krypton light upwards to as far as we can push the trigonometric method using parallax. In round figures this ranges from 10^{-8} metres (100Å) to 10^{18} metres (about 100 light-years); a compass of some twenty-six orders of magnitude, whose extremes are the size of a molecule of DNA and the distance to near-by stars.

The further we go outside this range the more uncertain determinations of length become, and the more reliant on the correctness of our theories they tend to be. In astronomy we can push inferences of distance out to 10^{22}m (10^6 light-years) by a fairly convincing method involving variable stars. This takes us beyond the dimensions of our own galaxy (10^{21}m) out to near-by galaxies. Further inferences and plausible arguments push distances out to the very edge of the universe, 10^{25}m (10^9 light-years) away. But at these distances one accepts the most convincing estimate as a working figure, ready to drop it whenever a more compelling one comes along, knowing that a new discovery or a new theory may arise at any time and require a change of all distances by a factor of ten.

At the other pole, measurements in the atomic régime can be pushed down to 10^{-10}m (1Å) by using X-rays and the wavelike properties of electrons. First, the wavelengths have to be defined. For both X-rays and electrons the phenomenon of diffraction through a perfect crystal provides a convenient measure; one which relies on the rather solid foundation of atomic theory, crystal structure, density measurements and classical wave theory. The determination of distances of the order of 1Å, whether by X-rays or by electron microscopes, is nowadays as familiar as using a metre rod to measure the length of a piece of string. As familiar – but, one should add, a hundred times more difficult to carry out.

Pushing down below the atomic scale demands increased resolution and this entails the use of shorter and shorter wavelengths. Thinking of the structure of the atom one might expect that the next scale of distance was determined by the size of the elementary particles. But here we have to be careful.

Let us ask a simple question – how big is an electron? Because of the wave-nature of the electron this is as valid a question as asking how big a photon is. It is obviously not such a simple question after all. Firstly, the photon associated with a given wavelength of light must be thought of as occupying the whole region of the wave-packet, which could be any size. The *minimum* size of a wave-packet is of the order of a cubic wavelength so, in a sense, we can regard the photon to be of those dimensions. The photon, unlike the billiard ball, does not have a hard and fast size. In the same way the electron, being a wave, cannot be smaller than its wavelength. Since the latter shortens as the electron speeds up, there is no one unambiguous size. The size of an elementary particle can-

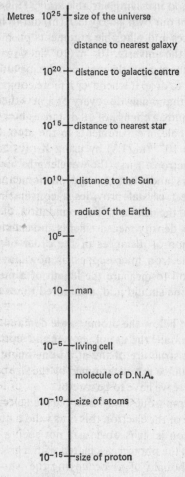

Figure 13. The span of space.

not be meaningfully separated as a concept from the dynamic quantities energy and momentum.

It becomes plain that at this level distance cannot be measured directly, and indeed becomes something of an irrelevancy. The important parameter is momentum, a dynamic quantity which can be more directly measured. Although it is related to a distance via the action quantum, the relationship is not usually invoked in describing physics at nuclear and sub-nuclear level. In an analogous way we find that energy replaces time, as we shall see. At the sub-atomic level we are usually quite happy to work entirely with the concepts of energy and momentum and to ignore periods and distances entirely.

Before we leave this topic let us refer to a length many times smaller than any already mentioned. It is a purely theoretical one, formed out of fundamental constants, namely the gravitational constant G, Planck's constant h and the velocity of light c. This length is $(Gh/c^3)^{\frac{1}{2}}$ and has a value of about 10^{-35}m. A photon with this wavelength would have a truly colossal energy. No one knows for sure the significance, if any, of this length. Perhaps it is a hint of how far we still have to go. As an indication, our largest planned accelerators are to produce particles of an energy which hardly begins to match even the energies of the most energetic particles in cosmic rays.

Finally, it is worth emphasizing the tremendous range of distances which presents itself. Perhaps the shortest distance that carries significance is the so-called Compton wavelength of the proton, 10^{-15}m, which is the size a proton appears to have when 'observed' with high energy γ-rays. (The Compton wavelength of the electron, incidentally, is much bigger, being about 10^{-12}m.) The longest is the radius of the universe, 10^{25}m. The ratio is the enormous number 10^{40}. Our measurements of length have to be good enough to cover forty powers of ten (Figure 13).

CHAPTER 4

TIME

To choose time is to save time.

Francis Bacon: '*Of Discourse*'

TIME is altogether a more elusive affair. We have a direct appreciation of events happening in a certain order, one following another. It does not matter whether the happenings are outside or inside our bodies, we can still order them into a series which recognizes earlier and later. If we experienced nothing at all then time would have no meaning. Time and events are indissolubly linked.

To measure time we need a reference series of events which we can label 1, 2, 3, etc. We can then judge that such-and-such an event occurred just as reference-event forty-four happened and so obtain the 'time' for that event. We ought not to use directly appreciated events inside our own bodies, since it would be nice to communicate time measurements to one another, and no two people could agree with any accuracy on their own personal time sense. So we need to decide upon a public reference series of events, a tram-line for time or, in other words, a clock. As far as we can see, no more than one degree of freedom is involved, so time is one-dimensional. If it turns out that some times are connected to other times by a different tram-line of events, then we will have to contemplate multi-dimensional time, but so far we have survived happily with a single reference series of events, in spite of ghosts, telepathy and H. G. Wells.

If all that we needed to provide a measure of time were a tram-line of events then almost anything would serve. Imagine a beetle clock. We put a beetle on a piece of graph paper and let it wander about. It ticks off a beetle second every time it crosses a line. In other words, an event is when the beetle moves from one square on the graph paper to an adjacent one. This is a truly ridiculous clock, but why?

One thing that is wrong is that we have to keep looking and counting beetle seconds. If we go away and come back again, we cannot tell how many beetle seconds have passed. A better arrangement is to cut a circular track out of paper and divide it up into equal segments, which can then be labelled 1, 2, 3, etc. (Figure 14). Now if the beetle is in segment 21 when we leave and in segment 54 when we come back we know that 33 b.s. have elapsed.

But what if the beetle is in segment ten when we return? Has time run backwards? Our intuition informs us that this is not the case. Or alternatively the beetle may have gone asleep on segment twenty-one. Has time stopped? We feel not. So in spite of having a tram-line of events the beetle clock just will not do. It offends our sense of the flow of time. Compared with that sense, beetle time is erratic and even reversible. We cannot believe that it would provide a simple framework for relating other events.

The worst thing about the beetle clock is the random walk of the beetle. A much more satisfactory series of events is provided by the lily. After a while, a lily on a pond will divide into two (Figure 14). The two will then divide into four, the four into eight, and so on. Here is the basis of a clock. To tell lily time we just have to count the lilies. Suppose there are thirty-two when we leave and 256 when we return, then 224 lily seconds have gone by. The nature of the lily clock is not erratic; the time is readily ascertained; there is no stopping and starting and reversing. But it will not do, even so.

Suppose we go away again for what feels to us the same length of time. When we return, the clock says 2,048, a lapse of 1,792 lily seconds. What we felt was about 224 l.s. turned out to be 1,792 l.s. Time had passed more quickly than we had thought. The discrepancy gets worse as lily time goes on. Everything seems to slow and become increasingly inert. We appear to live a vast *rallentando*. Once again, a perfectly good tram-line event proves to be offensive to our intuition of how time ought to be measured. The number of lily seconds in our intuitive second increases exponentially. We will find it easier to find a clock that does not offend our perception of time than work with clocks like the lily clock. The decision seems an arbitrary one, justified on the grounds of simplicity. But can we ever be sure that the clock we choose

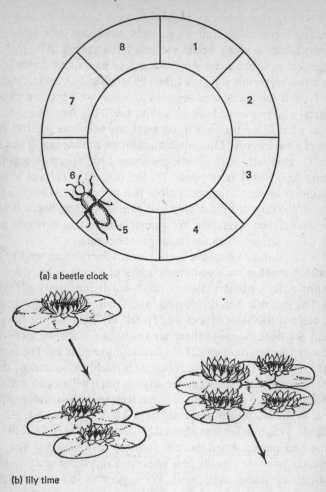

(a) a beetle clock

(b) lily time

Figure 14. Beetle clock and lily time.

has no gradual *rallentando* or *accelerando* built in to its nature?

For a long time our choice of clock was the earth itself. The series of events was provided by the rotation about its own axis, the basic event being the sun reaching its highest point in the sky at noon each day. Here was a clock which was so much a part of our direct experience, and indeed our very nature as human beings, that it could not possibly offend our intuitive feelings for time. But even so, small non-uniformities, irregularities, and long-term changes of motion had to be allowed for, before an acceptable unit of time could be provided. Thus the mean solar second was defined as the fraction 1/86,400 of a mean solar day, which was an average taken over a so-called tropical year, the time between successive vernal equinoxes. To take account of long-term changes, a particular year, namely 1900 A.D., was chosen as a standard. Nevertheless, the earth is as unique as a beetle, and, though not as unpredictable, it has its idiosyncrasies, enough to make it eventually as unsuitable for the purposes of defining the second as it was found to be for defining the metre.

Electromagnetic radiation once more provides the standard, this time the radiation emitted by caesium-133. The wave train provides a natural series of events – the wave crests (or any other given phase of the wave) – and the second is defined as the duration of 9, 192, 631, 770 periods of the caesium radiation. This emission, at a frequency of about 9 GHz (wavelength about 3 cm) is in the microwave region of the electromagnetic spectrum – the part used by radar. Sub-standards in the shape of electronic counters can easily be made, which can count events at frequencies up to a few hundred MHz. Direct measurements of time intervals of 10^{-10}s (100 picoseconds) are possible with sampling oscilloscopes, and electronic spectrum analysers measure frequencies of electromagnetic waves up to 40 GHz. The use of solid state properties is capable of pushing the direct measurement of frequency well into the infrared region of the spectrum. The limit is set by the time it takes for electrons in solids to respond measurably to electromagnetic radiation. This time varies from solid to solid, but a value of the order of 10^{-14}s is not untypical. Ever since electronic devices were invented, their response time has steadily decreased from milliseconds to microseconds and then to nanoseconds. Only recently have picoseconds entered significantly into the practical language of electronics (1 picosecond = 10^{-12}s). How long will it be before we hear the femtosecond (10^{-15}s) bandied about in a familiar way? Only, one might guess, when pure electronics is abandoned in favour of something faster, such as photonics – whatever that is.

But one may ask how we know that the response time of an electron in a solid is around 10^{-14}s if we cannot measure it directly. The answer is that we infer this time on the grounds of solid-state theory, about which we feel fairly confident. The actual measurements which allow us to put a number to this time are not in the least like a time measurement. But the results, fed into a theoretical expression, yield a time. More accurately, one obtains a quantity for the solid, under the given conditions, which has the dimension of time.

In much the same way we deduce the frequency of visible light. No one has yet measured directly a frequency of the order of 10^{15}Hz yet no one doubts that this is the magnitude for visible light. The actual measurements are, however, of wavelength and velocity and the frequency is inferred from simple wave theory. Wave theory also allows us to infer a frequency of about 10^{18}Hz for an X-ray of wavelength in the Ångstrom range. In this sense, times of the order of 10^{-18}s can be 'measured'.

Time at shorter intervals than this becomes increasingly irrelevant in practice, for the same reasons that length does at the sub-atomic level. If we want to measure the period between wave-crests in a γ-ray in practice we measure the energy of the γ-ray photon, which is comparatively straightforward. γ-ray photons are quantum particles, each carrying the quantum of action, h, equal to the product of its energy and wave-period. The latter can then be computed, knowing h and the energy. But it is energy, the quantity directly measured, which assumes practical significance, and the characteristic time associated with the γ-ray becomes of secondary importance. In the case of a radio wave, of course, it is the period which is most easily measured, and it is the photon energy which has to be inferred. Energy and time are conjugate quantities, joined together by the existence of the quantum of action. In this sense, short periods between events always imply large energies.

The smallest inferred time in physics is that required for light to traverse 10^{-35}m, the length obtained from fundamental constants introduced at the end of the previous chapter. This time is 10^{-43}s. Obviously it is a purely theoretical constant and so far it does not play a significant part in our understanding of nature. Perhaps the shortest times of real significance are the lifetimes of some of the short-lived elementary particles, which are measured by measuring lengths of tracks on a photographic plate. The grains of photographic emulsion cannot be made much less than 10^{-6}m across. A particle travelling near the speed of light must therefore live for at least about 10^{-14}s to produce a track. Statistical analysis of a large number of similar events can be made to yield significant results even when the lifetime is down to 10^{-16}s. All of which is as abstract as the procedure for assigning frequencies to X-rays. But to measure shorter lifetimes means eventually measuring energies and using the existence of the quantum of action. In this way lifetimes of particles as short as 10^{-23}s may be deduced. The larger the energy available the shorter can the lifetime be. Increasing the size of particle accelerators pushes our probing not only into smaller and smaller distances but also shorter and shorter time intervals.

At the other end of the scale clocks are needed to measure extremely long time intervals. We would like to know the age of dwellings uncovered by archaeological exploration, the age of rocks exposed by the geologist, the age of the earth itself, and finally the age of the universe. One of the most successful clocks which has been used to determine ages of things on earth, including meteorites which have survived the fall through the earth's atmosphere, is the radioactive atom. The radioactivity clock is like the lily clock in reverse. If we have a large amount of a given radioactive element then half of it will decay into some other element in a definite time, known as the half-life. Half-lives of naturally occurring radioactive atoms range from seconds to thousands of millions of years. If one is lucky enough to find radioactive atoms in whatever one wants to find the age of, then careful measurements and a knowledge of the half-life can indicate the amount of time which has passed since the atoms were enclosed within the object. In this way the origin of the earth and solar system is put at $4 \cdot 5 \times 10^9$ years ago.

To determine times on the cosmic scale we have no option but to use light itself as a clock. This requires that we know how far it is to stars and galaxies, and as mentioned in the previous chapter, the measurement of astronomical distances is not easy and straightforward. If we do know the distance to a star then a knowledge of the all-important velocity of light will tell us how long the light has been travelling. This intimate relationship between distance and time is embodied in the astronomical unit of distance: the light-year – the distance light travels in one year. The discovery that the more distant a galaxy was, the faster it was moving away from us, leads us to assume that this was caused by a vast explosion. In the 'big bang' theory this primeval explosion took place some 10^{10} years ago (Figure 15).

Here in the case of time, as in the case of space, the span which presents itself for measurement is enormous. To go from the shortest, significant time of 10^{-23}s to the longest 10^{17}s (10^{10} years) requires a factor 10^{40}. The same factor was found for distance – not surprisingly, since the velocity of light connected these time and distance measurements in a straightforward way. This role of light in connecting space and time is explored in greater depth in the next chapter, where it is shown that, in fact, the connection is much more intimate than one suspects at first.

The standard of length and the standard of time are both determined by electromagnetic waves. Having these standards we can determine the velocity of light, which is a fundamental constant of electromagnetic theory. This turns out to be $2 \cdot 997925 \times 10^8$ metres per second, which is accurate to 1 part in 10^6. Many physicists believe that a better system of definitions would be to adopt the velocity of light as a standard, instead of defining a standard of length. The metre would then be defined in terms of the time taken for light to travel that distance. This approach has the advantage that it corresponds to how distances are actually measured when there is no possibility of laying down a ruler, being much simpler than the parallax method. If we shoot a laser pulse at the moon and time the detection of the reflection, we can work out the distance very easily. All that is needed to map out the positions of objects is an apparatus which emits pulses of light,

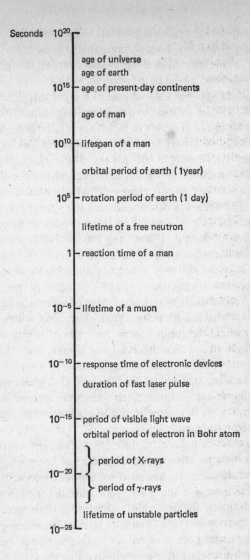

Seconds

- 10^{20}
- age of universe
- age of earth
- 10^{15} — age of present-day continents
- age of man
- 10^{10} — lifespan of a man
- orbital period of earth (1 year)
- 10^5 — rotation period of earth (1 day)
- lifetime of a free neutron
- 1 — reaction time of a man
- 10^{-5} — lifetime of a muon
- 10^{-10} — response time of electronic devices
- duration of fast laser pulse
- 10^{-15} — period of visible light wave
- orbital period of electron in Bohr atom
- 10^{-20} — period of X-rays
- period of γ-rays
- lifetime of unstable particles
- 10^{-25}

Figure 15. The span of time.

61

another to detect the reflected signal, and a clock standardized to the caesium radiation. Everything can be done from one place. Moreover, as mentioned in the previous chapter, straight lines can be usefully defined by the path followed by light.

The practical snag to this scheme at the present time is that the velocity of light, as defined by the prevailing standards, is known only to an accuracy of 1 part in 10^6. This is to be compared with a reproducibility of 1 part in 10^8 for the standard of length and a reproducibility of 1 part in 10^{12} for the atomic clock. If we defined a new metre in terms of the figure quoted above for the velocity of light, there would be too great a disparity with the old standard of length. The snag will disappear once the velocity is determined with an accuracy limited purely by the reproducibility of the existing standard of length. (This snag has probably already disappeared – the latest quotation of uncertainty is 0.004 in 10^6.)

The main point however is not concerned with standards; it is the choice of electromagnetic radiation to quantify space and time. What has been done is to define a frame of reference based on the interaction of charged particles, usually electrons, with light. These electromagnetic interactions form the series of events which serve to define time. In practice this is a good choice, since the vast bulk of everyday science is dominated by the laws of electromagnetism. But it does mean, and we should realize it, that phenomena which have nothing to do with the electromagnetic interaction, such as gravitation, the weak interactions and nuclear events, are going to be described within an alien framework. It might be that, such is the hidden unity of nature, no difficulties arise, and the framework only appears to be alien because we do not know enough about the world. It is generally assumed so, if only to avoid the complication of envisaging a space and time unique for each type of interaction. Imagine the growth of complexity, if we were forced to work with gravitation time, neutrino time, electromagnetism time, and nuclear time in order to reduce problems to manageable proportions. Nevertheless, it is a possibility, like the introduction of a fourth space dimension, which has to be taken out of the attic and dusted every now and again.

In spite of the strong impression of time as something flowing ever onward, there is nothing in the laws of motion, be they mechanical or electromagnetic, which differentiates between time running forward or backward. If a film were made of the collision of two particles and run backwards, the laws of physics would describe the reversed collision just as well as the original one. Sometimes this is used quite deliberately in theoretical descriptions. Suppose an electron and its anti-particle, the positron, move together and annihilate one another, producing γ-rays. This may quite happily be described in terms of an electron coming along, emitting γ-rays and then moving backwards in time, the latter being equivalent to the anti-particle moving forward in time. But this is just a trick played with the mathematics of the process and corresponds to nothing real. Nevertheless it is odd that the laws of interactions are independent of time reversal.

It is only in the realm of large numbers of particle that an 'arrow of time' is evident. A statistical quantity can be defined for a system which in all reactions either remains the same or increases. This quantity is the entropy, which is closely associated with degree of randomness, and the law describing its behaviour is the second law of thermodynamics. It is very much a statistical law. Parts of the system may actually reduce their entropy. Many writers refer to life itself as an example of entropy decreasing. But, overall, the sum of entropy always increases with every reaction which takes place. Things get more disordered on balance as time goes on. Time always increases – so does entropy. We will return to this topic in a later chapter.

Perhaps we are short-sighted in finding an arrow only in thermodynamics. Surely we could equally well point in cosmology to the expanding universe. If time increases so does the distance between galaxies. Are either of these in any way connected intimately with time? If the universe were contracting would things be different? On the other hand perhaps our laws of interactions will turn out to be asymmetric with respect to time after all, and an arrow will be found in the realm of elementary particles. The answer is the banal one. Only time will tell.

CHAPTER 5

MOTION

The spirit of the time shall teach me speed.

Shakespeare: *King John*

PURELY static things are really of little conceptual interest to physics. They may be used to present a framework against which comparisons can be made, or noted as limiting cases, but they are essentially peripheral. The main concern is with change. The simplest change which an object can make is one in which it retains its identity but moves from one position to another. The fact that things do move already influences our conception of space, in that three dimensions are required, and it is the prime reason that the concept of time is necessary. Moreover motion appears as an essential ingredient in some of the things of physics, for example, mechanical waves and the 'spin' of an electron. But even more important, the full appreciation of motion, as we shall see, leads to a broadening and deepening of our understanding of space and time and a stimulating appreciation of a real world fantastically different from the everyday commonsense one.

If we look around us at things on the move, the complexity is appalling – a leaf flutters in the breeze, a kitten tumbles over on its back in mock defence, an orchestra tunes up. As ever, in physics, we look for the simple and one of the simplest motions is of a thing moving in a straight line. If concepts can be worked out in this case then maybe a complicated motion can be analysed into more or less simple components and understood in that way.

Imagine we have the problem of describing the motion of a train travelling along a straight piece of railway track. Not wishing to get involved with the 'internal' motion of the train, such as the wobbling of carriages, we concentrate on the movement of the front of the train along the track and ignore the rest. We therefore station ourselves at a point *A* alongside the track and note the time when the front of the train passes us. All this tells us is the time and that the train is moving. To say something quantitative about *how* it is moving requires an assistant with a clock identical to and synchronized with ours at a point B further down the track, who notes the time when the front of the train passes him. After measuring the distance AB we can say the simplest quantitative thing about the motion, namely, that the train travelled so many metres in so many seconds, not forgetting to add a note on direction – the train was travelling from A towards B. We have measured the velocity, a derived quantity possessing a magnitude – so many metres per second – and direction – motion directed in a straight line from A to B. Velocity is an example of a vector. What could be simpler than our method of measurement?

The answer is, very little. But we should be aware of a vital assumption that has been made concerning the question of synchronization. No problem exists in testing that both clocks keep the same time at a point A. But how do we know that our assistant's clock remains synchronized after he has carried it to B? For all we know, time at B may flow at a different rate. Of course, we can always *assume* that the clock remains synchronized. Before Poincaré and Einstein around 1900 everybody made exactly that assumption. Nowadays one cannot take any assumption seriously unless it can be tested in experiment. This is the only way to distinguish physics from metaphysics – the way things really work from the way we would like them to work. We would like the clocks to remain synchronized when separated because it seems very simple, but we must forego our preferences and devise an experiment to test whether that assumption is true or false.

The problem is to get information about the clock at B to A in order to compare the two readings. So our assistant at B notes the time on his clock and dashes along the line to A with the time at B firmly memorized. When he gets to A he reads the time on the clock and compares with his memory of the time at B. The clocks are no longer synchronized! The clock at B apparently is running slow. But then we realize that it took time to convey the information about clock B to A, so of course it seemed that the clock at B was running slow. Clearly what is required is an instantaneous transmission between the two clocks. Unfortunately, nobody knows any way of conveying information instantaneously.

Naturally, something better than an assistant running between *B* and *A* can be devised. He could use a motorbike, or better still he could shout the information and let sound waves do the travelling. Best of all he could use electromagnetic waves. In fact the simplest thing to do is for us to look through a telescope focused on the clock at *B*. The clock at *A* can then be compared directly with the image of the clock at *B*. Even so an accurate comparison will reveal that the clock at *B* always shows an earlier time than that on the spot at *A*. Happily the difference in times remains constant provided the clocks are not moved closer together or farther apart. We can understand this difference by employing the concept that carrying information by light takes time. It took time for our assistant to convey information, and in the same way it takes time for light scattered from the clock at *B* to reach our eyes through the telescope at *B*. The difference in time-lag is one of degree only. With light the time-lag is small, but it is finite.

If this time-lag were known we could tell whether the clocks were still synchronized. All we would do is subtract that time-lag from the time registered by the clock at *A* and compare this earlier time with the image of the clock at *B* seen through the telescope. If the times are identical then the clocks are synchronized. The problem is therefore solved. All we have to do is to measure the velocity of light and so deduce the time-lag from the distance apart of *A* and *B*. Stop! Remember we are trying to measure the velocity of the train and we cannot do that without first sorting out this problem of synchronization. Without solving this problem we cannot measure the velocity of anything, including light.

We have hit a beautiful, basic impasse. To measure velocity we have to measure time at different points in space. To do this we need to test that clocks remain synchronized when moved from one point to another. To carry out this test we require a knowledge of the velocity of light. To measure velocity we have to measure time at different points in space ...

Problems like this will always occur whenever we split up a system into components and then try to understand the behaviour

of one component without reference to any of the others. In a
closed system like the universe it cannot be done. All we can do is
to use one bit of the world to investigate another bit. In this way
properties of one component are determined *relative* to another.

In our case we decide to adopt light as a reference, since it is the
fastest thing we know of and also it is an electromagnetic wave, and
electromagnetism is at the root of the structure of matter. We get
out of the impasse by taking the velocity of light to be one of the
fundamental constants of nature, whose value is given by electromagnetic theory, in terms of the electric and magnetic properties
of the vacuum. In principle we must *choose* a value for this constant, since we cannot attach a meaning to its measurement. In
practice, what is done is to assume that clocks remain synchronized in a given measurement and so derive a quantity which we
call the velocity of light. We can then use this value to calculate
the time-lag in any other situation, and so define what we mean by
synchronization.

So our way out of the impasse is essentially to assume a value
for the velocity of light and then define a convention for determining time at a distance from our standard clock in the way
already described. In fact, we can economize on assistants and
clocks and be very self-sufficient in our measurements if we can
generate and receive light pulses. Suppose we have such apparatus
together with our standard clock at A. Then if we can reflect a
light pulse from B we can not only define the time of an event at B
but also how far B is away from A (Figure 16). What we do is to
note the time a pulse leaves and the time it returns. We *define* the
time of reflection at B to be exactly midway between the measured
times at A, and we can define the distance between A and B as the
velocity of light multiplied by half the difference between the times.
The two readings at A therefore serve to define time and distance
associated with B. In this neat way we can map out the world,
measuring distances and defining time at a distance. We can now
measure the velocity of trains and anything else, and realize that
such measurements have a meaning only within the above convention.

In our attempt to measure motion we have discovered that there

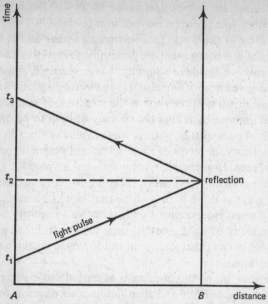

Figure16. Space-time diagram defining 'time-there' and distance. Clocks at A measure 'time-here': t_1, the time of emission, and t_3, the time of detection. 'Time-there', the time of reflection t_2 is defined thus:

$$t_2 = \tfrac{1}{2}(t_1 + t_3).$$

The distance between A and B is defined thus:

$$x = \frac{c}{2}(t_3 - t_1),$$

where c is a universal constant, known as the velocity of light. From these single definitions follow all the peculiar effects of special relativity.

are really two times. There is time-here, defined by our standard clock, and time-there, the time at a distant point, which eventually was defined in terms of time-here and the velocity of light. But what is time-here for us is time-there for another observer. Do these times agree? The answer is yes, provided the other observer is stationary with respect to us. London, New York and Sydney can all agree about time because they do not move with respect to one another (ignoring continental drift).

But what happens when the other observer is moving? Things are not as simple. To him it appears that *we* are moving and so he has to determine time-there at a place which keeps changing its position. He must, of course, use the same tools and the same convention as we use, otherwise we will not be talking in the same language. What an observer, moving with uniform velocity relative to us, measures is easily deduced (easy once Einstein showed us how to do it in his Special Theory of Relativity) (Table 2).

Table 2. Relativistic effects in moving objects

length	lengths along the direction of motion appear shortened by a factor $\sqrt{1-(v/c)^2}$.
contraction	lengths at right-angles to the direction of motion are unchanged.
time dilation	moving clocks run slow by a factor $\sqrt{1-(v/c)^2}$.

v = relative velocity, c = velocity of light.

Though easily deduced, his results appear totally bizarre. Einstein's Special Theory of Relativity says that he will observe our time to flow more slowly, and our objects will appear to be rotated. Equally, if we observe him and his travelling laboratory, we will see his clocks running slow and his objects twisted around. Part of the rotation of objects is due to a relativistic contraction of length in the direction of relative motion and the rest is due to the appearance of sides, which would be invisible at rest (Figure 17). For him, a moving thing appears to function in a framework in which the shortening of length is compensated by the dilation of time. We come to exactly the same conclusions about his world. If we both observe the same physical phenomenon, say the creation and decay of a muon, we will disagree about periods and lengths. If the muon is not moving very fast with respect to the observer he will measure a lifetime of about a microsecond. If the muon travels close to the speed of light with respect to us we will see a much longer lifetime. To us the muon travels perhaps ten kilometres between creation and decay. To the rapidly travelling

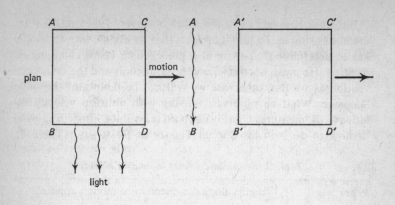

plan

motion

light

observer

(a) moving cube observed.

observer

(b) light emitted from edge A travels length of side as cube moves from A to A'. Light emitted from B' will arrive at observer at same time as light from A.

elevation

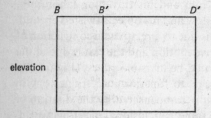

(c) instantaneous view. Observer sees trailing face. Without relativistic contraction cube appears distorted.

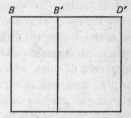

(d) physical contraction of length in direction of motion (B'D') makes cube appear rotated, but not distorted.

Figure 17. Appearance of rapidly moving cube.

observer the distance is much shorter. It is clear that our convention of measuring time-there and distance entails the corollary that such measures of time and space are purely relative to the observer.

Our convention seemed very common-sensical and simple when we introduced it. Why does it produce such odd effects when we consider uniformly moving observers? The oddness stems from the adoption by each observer of the same value for the velocity of light. On the one hand this adoption is entirely reasonable. The velocity of light appears in Maxwell's equations for electromagnetism as a fundamental physical constant. Furthermore, Newton pointed out the fact that uniform motion was the natural state of things when no forces were acting on them. Since nothing exists which can be identified as being at absolute rest, all uniform motion is relative. One uniform motion is as good as another. The laws of nature should not therefore depend upon relative velocity and hence the velocity of light must be the same fundamental constant to all uniformly moving observers. So the adoption of the same value for the velocity of light by all observers in order to map out their time and space is not only reasonable; it is inevitable.

On the other hand our experience tells us that if we are in a car travelling at 100 kilometres per hour (in short, km hr^{-1}) and we fling out of the back an apple core at a velocity of 20 km hr^{-1} with respect to the car, we know that the disapproving gentleman on the pavement will see the apple core travel past him at a speed of 80 km hr^{-1}. Relative to us the apple core travelled at a velocity of 20 km hr^{-1}: relative to him the speed was 80 km hr^{-1}. If the apple core were a photon, occupants of pavement and car would agree that it travelled at the velocity of light – the speed of the car does not affect matters. In this way photons are quite non-common-sensical objects. A photon travels with the speed of light relative to all observers no matter how fast they are going. Quite mad!

Do we abandon our convention, because it predicts funny things, or do we abandon our belief that the laws of physics should be invariant among uniformly moving observers? As ever, the only arbiter is experiment, and experiment shows unambiguously that our convention agrees with the way the things of physics behave. We have to accept the fact that just as things get more uncommonsensical at small lengths and short times and also at large distances and long times, so do things appear bizarrely unrelated to personal experience at high velocities.

The time and space that is familiar to us is just our own way of perceiving an underlying reality, which we have learnt to call the four-dimensional space-time continuum. Each observer 'sees' a particular way in which space-time splits up into a dimension of time and three dimensions of space. What human or mythical observers 'see' would be fairly irrelevant if it were different from what physical objects 'see'. But, in fact, elementary particles travelling close to the speed of light do behave as if they 'see' a carve-up of space-time which an observer travelling at the same speed would 'see'. Thus muons *do* live longer at high speeds. Experiment shows that lengths in the direction of motion *do* contract. If we were to travel near the speed of light we would live longer according to earth time (though we would not notice any difference). Thing *do* behave according to relativistic theory.

Though relativity mixes up time and space it does so in a precise and unambiguous way. The relationship between the time and space of one frame of reference and the time and space of another is described by down-to-earth, simple, algebraic equations – the Lorentz transformations. No change in physics results from hopping from one frame of reference to another – only the details.

If one observer sees an event at a place A and, later, another event at a place B, then the event at B could have been caused by the event at A if, and as far as we know, only if, a photon or something slower, could have travelled from A to B in the time between the events (Figure 18). If indeed a photon could have gone from A to B then all observers, no matter what their relative velocity, will agree that the event at A came before the event at B. Therefore all will agree about later and earlier when it comes to describing causes and effects, so physics remains unaffected. But if a photon could not have done the journey, some observers will be found who will see the event at B happen before the event at A. Past and future are no longer absolute for such events. But since the events could

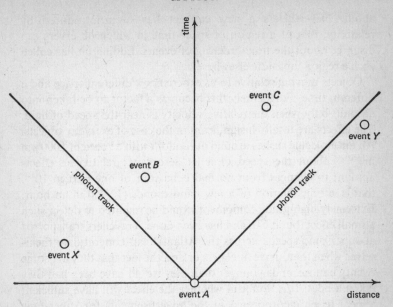

Figure 18. Absolute past and future and absolute elsewhere. Events B and C could have been caused by event A, since whatever may have travelled between A and B, and A and C, could do so at a speed lower than that of light. All observers agree that B and C occurred after A. All events lying in the region between the photon tracks occur in the absolute future stretching from A.

Events X and Y could not have been caused by event A, since nothing travels faster than light. Some observers will see X and Y occurring before A. Since neither X nor Y are physically connected to A, disagreement over time-ordering is irrelevant to physics. All events lying in the region below the photon tracks occur in the absolute elsewhere stretching from A.

not possibly affect one another (since the fastest transmission of interaction is light) physics remains indifferent to the disagreement. So, though at first sight it seems that the relativistic juggling with time and space could lead to a mad world, it turns out that the world, albeit distorted, remains quite sane. There exists still an absolute past and absolute future which all observers agree about. Nevertheless a new concept has been introduced by relativity, that of a region of space-time in which observers can disagree about the time-ordering of events. Eddington has called such a region 'absolute elsewhere'.

Objects moving relative to us experience a different space and a different time. Fortunately this becomes a factor to be taken into account only when the relative velocity is near the speed of light. The effects are totally insignificant in the case of everyday objects. An air molecule rushes around at random with a speed of 1000 km hr^{-1} – about the speed of a jet aeroplane; relativistic effects amount to changes from normal behaviour of one part in 10^{12}, that is, a time dilation of a few nanoseconds (10^{-9}s) in an hour. Extremely high-grade equipment would be required to detect such a small effect, but it can and has been done. Travellers transported at supersonic speeds across the Atlantic can congratulate themselves when they arrive in New York on the fact that they are some twenty nanoseconds younger than they would have been had they taken a boat. It is doubtful whether the effect will have sufficient appeal to be incorporated in the advertisements for supersonic flight.

What would happen if something was discovered which travelled faster than light? Everyone would be fascinated and delighted and there would be another revolution in physics. Light defines electromagnetic time and electromagnetic space in which as an inevitable (tautological) consequence Maxwell's equations are true. The new superluminary particles would, if used in the same way as light, define a totally new time and space, in which Maxwell's equations would change. The speed of light would, in this new time and space, no longer be a fundamental constant, the same for all uniformly moving frames of reference. Extra rest-mass energy in particles would be found. Indeed, theoreticians *do* play around with theoretical particles which travel faster than light – these hypothetical superluminaries are called *tachyons*, and they fit quite happily into the existing electromagnetic scheme of time and space provided their velocity *never falls below* that of light, But, as yet, nothing has been found which has this vital property.

Even if something *were* found, it would not mean we would

have to give up the basic idea of relativity, which is that the physics of the world is not discovered to be different in laboratories travelling with uniform velocity relative to one another. It seems to us that one uniform velocity is as good as another. If this were not so there would be an absolute hierarchy of uniform motion, incorporating the special case; a frame of reference which was at absolute rest. What absolute rest is nobody knows. We work on the assumption that something can be at rest only with respect to something else.

Uniform motion, that is motion with constant speed in a straight line, is of special significance in physics, as the foregoing discussion has amply illustrated. Before Newton, it was not realized that such motion was the natural state of things. It seemed necessary to account for the continuing flight of an arrow once it had left the bow. What kept it going? Newton discovered that the more productive question was – what stopped it going? Since Newton, the questions – what stops things going? and – what starts things going? have dominated the field. The apparently equally valid question – what keeps things going? has been shelved by attaching a label to it. We could have said that what keeps things going is magic, but to be more precise (since magic is a large field in itself) we say that what keeps things going is inertia. Some people nowadays have begun to think about the origin of inertia – more of this in later chapters – but the essence of physics is to do with interactions. And interactions are observed by seeing changes from uniform motion.

Having brooded on the significance of our measuring technique we can return to the railway track and start to quantify the motion of our train. We bounce a pulse of light at the front of the train noting the time the pulse is emitted and the time it returns. Those two measurements reveal, by our convention, position and time. Repeating the operation gives us a second position and a later time from which we deduce the velocity. Is it uniform? A third pair of measurements tell us a third position and time. From the second and third operations the velocity can again be deduced. If identical to the first velocity, the motion is uniform. If different, the motion is non-uniform and from the difference in velocities and time between the velocity determinations, the rate of change of velocity – the acceleration – can be deduced.

The simplest acceleration is a constant one. To test for this needs a fourth operation. If the change in velocity in a given time is constant then the acceleration is uniform. If not, a change in acceleration can be measured. The simplest *non-uniform* acceleration is one which is proportional to distance. To test for that requires a fifth operation – making ten readings in all. In practice, to improve accuracy, we would take many more readings than ten, employing automatic recording equipment – such as a fast movie-camera – to help us. But it is basic to realize that a measurement of velocity cannot be made unless at least four readings can be taken. Nor can the simplest non-uniform acceleration be confirmed this side of ten readings.

Moving away from the motion of our train to the motion of general things, we can identify four types of non-uniform motion which crop up time and again in our models of the universe (Figure 19). One is the case of uniform acceleration in a straight line, exhibited to a greater or lesser degree of approximation by cars, trains and aeroplanes. It is associated with a constant force applied in the direction of travel. A second type is the case of a uniform acceleration which is always directed towards a certain point in space, as occurs in the case of circular orbital motion – approximately that of the earth moving round the sun, or a stone whirled around at the end of a string. The earth is continually accelerating towards the sun, but this motion is continually offset by the tangential motion, so it gets no closer – a case of a change in the direction, but not the magnitude, of velocity constituting the acceleration. A third type is the case of an acceleration always directed to a certain point in space, but proportional to the distance from it. This results in what is called simple harmonic motion – the swing of a pendulum, the motion of matter in a sinusoidal mechanical wave, a mobile electron responding to the oscillating electric field of a sinusoidal radio wave.

The fourth type is rather different. It is the case of a transient interaction in which the motion changes from one sort of uniform motion to another, or one sort of simple non-uniform motion

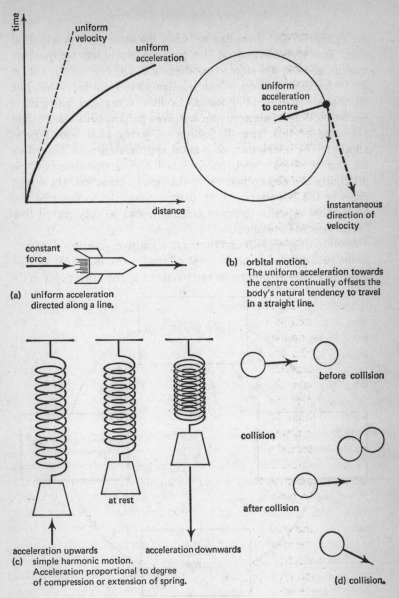

(a) uniform acceleration directed along a line.

(b) orbital motion.
The uniform acceleration towards the centre continually offsets the body's natural tendency to travel in a straight line.

(c) simple harmonic motion.
Acceleration proportional to degree of compression or extension of spring.

(d) collision.

Figure 19. Four types of non-uniform motion.

like types two and three to another of the same type, in a limited time. This type abounds in the quantum world and we speak of transitions from one state to another state. The emphasis is not on the transitional motion, which is often plainly unobservable, but on the initial and final states. Collisions between particles – electron–electron, electron–photon, even billiard balls – are prime examples of this type. Reflection of waves at a boundary is another. The interaction, to a good approximation, is limited to the region of space and time in which the particles are in close proximity to one another, or, in the case of reflection, the waves close to the boundary. Many theoreticians like to describe the three other types in terms of collisions, and so they regard this fourth type as fundamental.

Motion of types two and three are repetitive – a given situation keeps recurring after an interval of time – the oscillator swings through its centre point, the earth returns to the same point in its

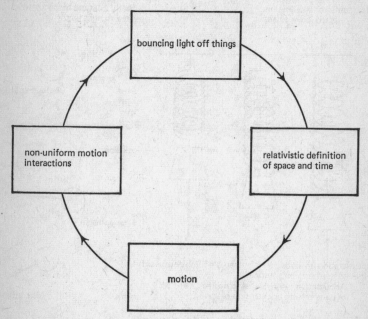

Figure 20. The interaction loop.

orbit once a year. The given situation in such cases is known as the *phase* of the motion and the time between repetitions of identical phase is known as the period, the reciprocal of which is termed the frequency. What is important about such motions is that they provide measures of time – the earth, the pendulum of a grand-father clock, the mechanical vibrations of a crystal of quartz, the electromagnetic vibration of caesium radiation. The continual repetition of phase beats out time.

The situation is therefore as follows. Non-uniform motion pro-vides our measure of time, which, with space, provides a frame-work in which other non-uniform motions are described. Space itself cannot be measured without bouncing light off things – non-uniform motion of the fourth type. The things of physics themselves depend upon non-uniform motion for identification. And the source of a non-uniform motion is an interaction.

Here is the conceptual loop of physics (Figure 20). Interactions define the things of physics. Interactions define space and time. The behaviour of the things of physics in space and time define interactions!

CHAPTER 6
ENERGY

Energy is Eternal Delight.
William Blake: *Marriage of Heaven and Hell*

PHYSICS, once it has decided on what simple things exist and how best to measure and describe their motion in a space-time framework, has the task of describing, in the simplest way possible, the interactions which occur between them. Looking for the simplest, in this context, means looking for the most generally applicable laws with the fewest concepts. The two vital concepts here are energy and momentum. These quantities, precise derivatives of our intuitions about work and force, stopping and starting, and so forth, find application in all interactions, whatever their origin – so much so that one may quite accurately describe the physics of interactions as the study of the exchange of energy and momentum. But at the root of the concept of momentum, and basic to the idea of energy of motion (kinetic energy), is another – the concept of mass.

To get at a meaning for mass which is somewhat more meaningful than the text-book metaphysical 'quantity of matter', let us go into our thought laboratory and conduct a few experiments with interactions. At this stage we will stick to macroscopic objects and deal entirely within classical, that is non-quantum, physics. Of the two types of interaction known to classical physics – gravitation and electromagnetism – we will choose the weaker, gravitation, to start with. Naturally, we have all the best apparatus with which to measure motion in the way already described. We also have means of counting precisely the number of atoms in any object. And, finally, we have magic screens which can be used to isolate the experiment from all other interactions.

We choose a large sphere made of platinum, count how many atoms it consists of, and float it in space. A single platinum atom is introduced into space some distance away from the ball and we observe that the atom accelerates towards the sphere and the sphere accelerates towards the atom. There is an interaction in the form of a mutual attraction. Quantitatively, the ratio of the accelerations is in inverse ratio to the numbers of atoms – the single atom accelerates much more rapidly than the ball. By some simple experiments we can tell that the accelerations are dependent upon position. We discover the interaction has a spherical spatial symmetry and its strength falls off as the inverse square of the distance apart (Figure 21). It exhibits no dependence on time.

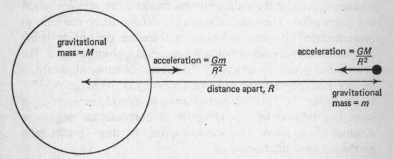

Figure 21. Gravitation. The gravitational constant, $G = 6\cdot670 \times 10^{-11}$ m^{3}Kg^{-1}s^{-2}.

But is the acceleration of the atom dependent on something in the atom, as well as something in the ball? We tack on a second atom of platinum and observe how the pair of platinum atoms move. The acceleration of the pair is exactly the same as when there was one atom! The acceleration of an object is therefore entirely the property of the ball and nothing to do with the object. We verify this by doubling the size of the sphere and observing a doubling of all accelerations.

We conclude that a platinum sphere is always surrounded by what we can call an *acceleration field*. Associated with every point in space there is an acceleration, which any object will experience if placed at that point.

Is the type of atom producing the field important? We discover that a ball made of an identical number of atoms of copper produces a weaker field, carbon an even weaker field, solid hydrogen weakest of all. Each atom has a characteristic quantity associated with it, which determines the strength of the field: a charge of acceleration-producing stuff. We call this quantity the *gravitational mass*.

To measure it, a standard is defined. In practice a chunk of platinum-iridium alloy is defined as having a mass of one kilogram (Kg). A Kg sphere of platinum-iridium will produce an acceleration at a given point of so many metres per second per second. To measure an unknown gravitational mass we replace the platinum-iridium sphere by the unknown and measure the acceleration at the given point. The ratio of masses is defined to be the ratio of accelerations. Gravitational mass is therefore defined as being proportional to the acceleration it induces at a given distance. The constant of proportionality, a fundamental quantity of nature, is the *gravitational constant* denoted always by G. Its value, 6.670×10^{-11} $m^3Kg^{-1}s^{-2}$, is the acceleration, in metres per second per second, produced by one kilogram of gravitational mass at a distance of one metre. This acceleration is very tiny – gravity is an extremely weak interaction.

It does not seem very weak when we fall, or try to get a spaceship to the moon, but compared with other types of interaction it is. A man strolls at a velocity of about one metre per second. A gentle acceleration of one metre per second per second takes him into a jog trot. The effort entailed is small. The acceleration at the start of a hundred metre sprint may reach 2 or 3 ms^{-2}. It takes $10^{24}Kg$ of earth to produce that order of acceleration at the surface due to gravity. An athlete, using what is essentially the electromagnetic interaction, achieves it with a few kilograms of muscle.

Though weak, this acceleration field we call gravity is ubiquitous. It cannot be switched off or neutralized. It inhabits the universe as firmly as space and time. Because both negative and positive electric charges exist, the strong electric interaction can be exactly eliminated by having equal amounts of charge of either sign. Gravitational mass, as far as we know, has one sign only and always gives rise to an attraction.

Yet much of the behaviour of matter in everyday things is not associated with gravitation at all. The principal interaction is that of electromagnetism. All mechanical, as well as electrical and magnetic, properties of matter are at base due to this interaction. It therefore behoves us to enter our thought laboratory again to investigate the basic elements of this phenomenon. No one will be startled when we emerge with the concept of electric charge and the electromagnetic equivalent of the gravitational constant, but it is surprising that in addition we have to bring out the quite different concept of *inertial mass*. In discovering the basic gravitational interaction, we could do without the concept of inertia. In discovering the basic electrical interaction we are forced to invent it.

We will work with individual electrons and protons. Let us begin by putting a countable number of electrons on our platinum sphere and observe the motion of a single test electron which is placed on its own, some distance away. We discover it accelerates away from the sphere, and we note that, compared with the previous accelerations in connection with gravitation, this acceleration is enormous. Otherwise the symmetry and inverse square law is identical to gravity.

Is the field surrounding the charged sphere an acceleration field, independent of object accelerated, as it was in the case of gravity? To test this, we would like to repeat the experiment with two electrons stuck together but – we cannot, they would fly apart. We abandon this experiment for the moment, and ask instead – do protons have the same effect as electrons? The electrons on the sphere are replaced by an equal number of protons. The test electron exhibits the same size acceleration, but this time towards the sphere. If we double the number of protons the acceleration doubles. From these experiments we conceive that each proton has an amount of something which produces acceleration – we call it *electric charge* – and that each electron has an equal amount producing acceleration in the opposite direction. By convention we say a proton has a positive charge and an electron has a negative charge. Accurate measurements show that these charges are equal in magnitude to an accuracy of one part in 10^{20}!

A new feature appears when the test electron is replaced by a proton. All the accelerations are reduced by a factor of nearly two thousand (1,836). The acceleration field surrounding the sphere is not independent of the object accelerated – so we need not think about how to stick two electrons together after all. A charged sphere produces an entirely different acceleration field for electrons and protons, quite apart from effects due to the signs of charges involved. Protons 'see' a weaker acceleration field than electrons. There is something else apart from sign of charge which makes a proton different from an electron. We call that something *inertial mass*, a quantity inversely proportional to the acceleration observed. Protons have 1,836 times more inertial mass than electrons.

The electric field is not, as it is in the case of gravity, an acceleration field, because what acceleration is produced depends upon the thing being accelerated. We call it a *force field*; defining force in terms of the acceleration of a unit of inertial mass. The force is then seen to be proportional to the product of the charges and inversely proportional to the square of the distance apart (Figure 22).

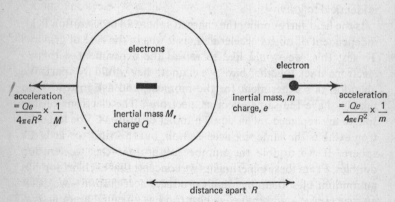

Figure 22. Electrostatic interaction. Like charges repel, unlike attract. The force is proportional to the product of the charges and inversely proportional to the distance apart squared. The constant of proportionality is denoted $\frac{1}{4\pi\varepsilon_0}$, and ε_0 is called the permittivity of the vacuum, a fundamental constant, equal to $8\cdot854 \times 10^{-12}$ farads per metre. The acceleration is the force divided by the inertial mass.

To quantify comparisons a unit of charge is required. The simplest would be the charge on the electron but, historically, it was defined quite differently. The unit is the coulomb (C), defined as the amount of charge passing through an electrolytic cell which will deposit $1 \cdot 118 \times 10^{-6}$Kg of silver from an aqueous solution of silver nitrate. The electron's charge is $1 \cdot 6 \times 10^{-19}$ coulomb – also the magnitude of the charge on the proton. The coulomb, like the kilogram, metre and second, is a good-sized practical unit.

A unit of inertial mass is also needed. To arrive at a unit of inertial mass imagine the experiment extended to cover α-particles (two protons and two neutrons combined) and charged atoms (ions) of all kinds. A study of the accelerations in each case reveals a remarkable fact. The ratio of inertial masses for the charged atoms, as required by the accelerations, is, allowing a fixed amount for the electronic mass, identical to the ratio of gravitational masses for the same atoms. In other words, the gravitational mass, a measure of the power accelerating other bodies, is identical to the inertial mass, a measure of the acceleration the body itself experiences under non-gravitational (at least electromagnetic) forces. The unit of inertial mass is therefore the kilogram chunk of platinum-iridium. The equality of the two masses, observed to accuracy of better than one part in 10^{11}, is an intriguing fact, providing a solid link between gravitation and electromagnetism. What the significance of this fact is, remains a mystery.

The way that the concept of inertial mass appears in our imaginary experiment is because the acceleration of the proton is found to have a different magnitude from that of the electron. This is not the only factor at the root of the situation. Perhaps a more compelling reason for the introduction of inertial mass is provided by the existence of equal and opposite charges. Imagine performing the experiment with a positron (anti-electron) instead of a proton. The *magnitude* of the acceleration would be identical to that of the electron. The charged sphere would then have surrounding it a field determining the size of acceleration – as in the gravitational case – but the direction is determined by the charge of the test object. At this stage no inertial mass is required. But now repeat the experiment with positronium, an atom consisting of a positron and an electron, mutually bound. Positronium, being overall neutral, naturally does not suffer any acceleration whatsoever (apart from a small one due to polarization effects, which can be safely ignored in this context). Suppose we can restore the acceleration by adding another electron to form a positronium negative ion. Compared with a single electron, the acceleration would be reduced by a factor of three. We therefore have to introduce inertial mass in order to describe the acceleration of a body in which the neutralization of one sign of charge by the other takes place. Does this explain why the proton is so heavy? We have not the faintest idea.

The electric interaction is not the only one in electromagnetism. There is also a force between moving charges, which we term the magnetic interaction; and there is also the production of electromagnetic waves by accelerating charges (Figure 23). The electric interaction however, is the principal one operating in mechanical events – albeit at the microscopic level. The strength of materials, the elasticity of solids, the atomic forces which come into play in mechanical collisions of all kinds, are all, at base, electrostatic in origin. Mechanics, therefore, cannot avoid including inertial mass in its concepts. It enters even in 'purely' gravitational problems, if the problem is about the gravitational force on *extended* bodies, since the various parts of an extended body are held together by electromagnetic forces.

When a moving billiard ball strikes a stationary one, it shares its motion with it. There is something which the moving ball had before the collision which the stationary one did not have. That 'something' is, in some sense, an amount of motion, a quantity that moving objects possess and stationary ones do not.

But surely a measure of this is the body's velocity. Here we already have a quantity to serve as the 'amount of motion'. Specify the speed and direction and we specify the motion, a magnitude with direction, in a word, a vector. This is the simplest idea, but it will not do nevertheless. A fast-moving tennis ball can easily be stopped by catching it, but it is much more difficult to stop a slow-moving motor-car. A heavy, steel ball-bearing has much greater power to dissipate a group of ordinary glass marbles than has a

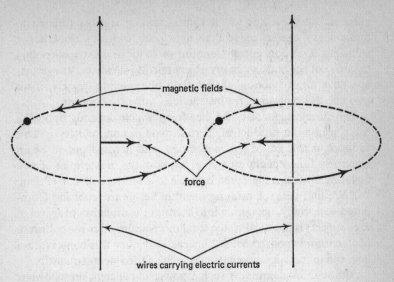

(a) magnetic interaction. Moving charges produce magnetic fields.
Magnetic fields exert a force on moving charges. The two wires
attract one another.

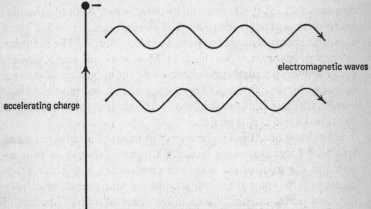

(b) accelerating (or decelerating) charges radiate electromagnetic waves.

Figure 23. Electromagnetic interactions.

glass marble travelling at the same speed. It is more difficult to make heavy objects move than light ones. Velocity alone is not sufficient. Our concept of 'amount of motion' must incorporate the mass of the body. A heavy object moving slowly may nevertheless have more 'motion' than a light one moving quickly, if we measure this motion by the interactions with other objects.

The simplest quantity we can form which depends upon the mass is just the product of inertial mass m and velocity v. This product, mv, is called the *momentum*, and it turns out to be an extremely fundamental quantity. Suppose the momentum of two colliding objects is measured before and after the collision (Figure 24). We find that the total momentum before the collision (computed vectorially, i.e. taking into account the direction of travel of each object) is identical to the total momentum after the collision. Momentum is conserved. All interactions obey this conservation law, and that is why momentum is such a fundamental quantity.

Although momentum is such a powerful concept, on its own it is not sufficient to characterize a motion completely. Think of stopping that slow-moving car with muscle power. It takes a lot of sweat. It takes a lot of sweat independent of which direction the car is moving (provided it is on a flat road). Momentum is a directional quantity, the amount of sweat is not. A lot of muscular energy is expended in destroying the motion. Equally well, a lot of muscular energy is needed to push a car into motion. The everyday concept of energy is applicable here. Motion entails a directionless (technical term, scalar) quantity, an amount of energy. A moving body can be thought to possess energy of motion, termed *kinetic energy*, which, anthropomorphically, is a measure of the muscular sweat produced in stopping it.

How do we quantify kinetic energy in terms of inertial mass and velocity? There is no easy intuitive answer. It should be proportional to the mass of the body like momentum, and it should increase with the velocity in some way, but beyond that we must infer the most useful relationship from the study of a large number of collisions. It turns out that the quantity of significance is the product of the mass and half the square of the velocity – in symbols $\frac{1}{2}mv^2$. And the significance resides in the fact that this quantity,

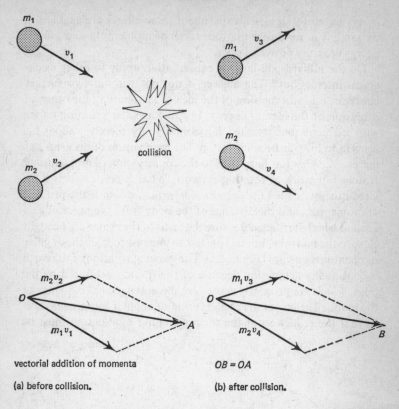

Figure 24. Conservation of momentum. In addition, kinetic energy is conserved:

$$\tfrac{1}{2}m_1v^2 + \tfrac{1}{2}m_2v_2^2 = \tfrac{1}{2}m_1v_3^2 + \tfrac{1}{2}m_2v_4^2$$

like momentum, is conserved in billiard-ball collisions. The total kinetic energy before impact equals the total after. The laws of the conservation of momentum and kinetic energy enable us to predict the outcome of all manner of billiard-ball collisions. What small discrepancies might occur can be tracked to the production of heat (internal motion) on impact, to air-resistance and to friction. But in any carefully designed experiment these effects are small, and we are left in no doubt that our ideas of momentum and kinetic energy are truly powerful ones.

Do the billiard-ball laws of conservation apply to other mechanical interactions? The answer is no, not without two further concepts, one an extension of the idea of momentum, the other an extension of the idea of energy. Imagine a skater spinning on ice with his arms outstretched horizontally. By merely pulling his arms in to his side he spins faster. The momentum of his arms has increased. What has happened to the conservation of momentum? It does not work in rotating systems. What is conserved is a related quantity called the *angular momentum*, which is the product of momentum and the distance of the body from the point of axis around which it rotates. Because the arms of the skater are brought close to the body the momentum has to increase to keep the angular momentum constant (Figure 25). The speed of rotation of the earth is continually decreasing because of the friction produced by the tides, which are primarily caused by the gravitational influence of the moon. The law of the conservation of angular momentum tells us that the reduction of the earth's angular momentum must be

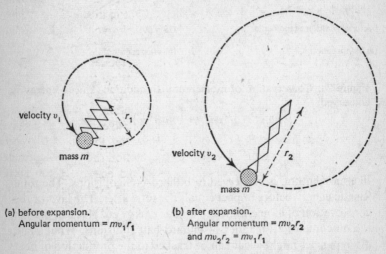

(a) before expansion.
   Angular momentum = mv_1r_1

(b) after expansion.
   Angular momentum = mv_2r_2
   and $mv_2r_2 = mv_1r_1$

Figure 25. Conservation of angular momentum.

balanced by an increase of the angular momentum of the moon in its orbit around the earth. The moon must be moving away from us (but the effect is tiny). If momentum associated with motion in a straight line is called linear momentum, then we can summarize by saying that in any mechanical interaction the total linear momentum and the total angular momentum separately remain constant.

Let us now develop our concept of energy. It is immediately obvious that the kinetic energy of any chunk of matter we hold at arm's length and drop does not remain constant. The body accelerates and increases its kinetic energy as it falls. But it does so under a strict gravitational law; the kinetic energy it acquires turns out to be directly proportional to the distance through which it drops. Kinetic energy comes from moving from one point in the gravitational field to another. There is something that remains constant. It is the sum of the kinetic energy and a quantity depending on position in the field. This quantity is a characteristic energy associated with every point and known as *potential energy*. In the unimpeded motion of a body from one point to another the sum of potential and kinetic energy remains constant. Before release, the body has zero kinetic energy and a certain amount of potential energy defined by its position in the field. After release, potential energy is lost and an equal amount of kinetic energy is gained. The total energy remains constant (Figure 26).

The concept of potential energy applies to motion of charged particles in electric fields. The electrical term voltage is just the potential energy of a particle carrying unit charge. Voltage measures the ability to produce motion of electric charge. The idea also finds application as the stored energy in a compressed or extended spring. A body hanging on the end of a spring can vibrate up and down. In such a motion the body acquires kinetic energy from the potential or stored energy of the compressed spring and subsequently gives it up in extending the spring. This exchange of kinetic and potential energy repeats itself in each cycle of the oscillation, but at all times the sum remains constant. Potential energy is an idea we have to use whenever there are forces acting.

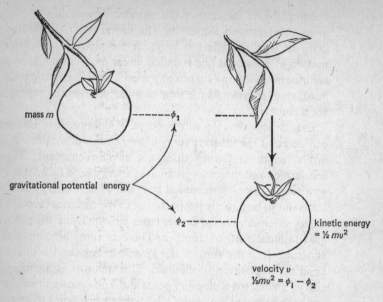

Figure 26. Transformation of potential energy into kinetic energy. Gravitational potential energy, $\phi = \dfrac{GMm}{R}$. In our picture M is the mass of the earth and R is the distance of the apple from the centre of the earth.

If we were not so used to things dropping when we let go of them, it would seem quite magical that motion appeared out of nothing. That bodies acquire motion from a palpable push is utterly familiar, with a familiarity that stretches back to infanthood. But things in a force field start to move without anything visible pushing them. Pure magic, but we have talked ourselves into behaving as though such things are perfectly understandable. Space surrounding a charged sphere acquires the extraordinary property to move charged things. We speak about its force as though it were a material push of one body in contact with another. We assign to every position a potential energy, being a potential for producing motion. We can describe what happens quite accurately and we think we understand. But, really, we do not. The invisible influences of gravitation and electromagnetic fields remain magic; describable, but nevertheless implacable, non-human, alien, magic. Potential energy is a measure of the strength of this magic.

Wherever forces are acting, momentum is no longer a conserved quantity. The definition of force as mass times acceleration can be put in another way. Force is the rate of change of momentum with time. A falling body acquires momentum because, we say, it experiences the force of gravity. It also acquires kinetic energy from the change in gravitational potential energy. Force and potential energy are the concepts which link the field to physical motion of a body. But, even so, our description of motion retains the memory of the magic: both momentum and kinetic energy contain in their definitions that mysterious quantity, the inertial mass, which, if you remember, is a concept which we had to introduce to understand the motion of charged particles in an electric field. And reflect that the familar push of one body on another has, at base, an electromagnetic origin, so the magic really inhabits the most commonplace events.

And what could be more commonplace than the bill we have to pay for energy consumed in the home? The magic may wear a little thin at this point, but it is nevertheless there, defining the unit of energy. In terms of kinetic energy the unit is just the energy an inertial mass of two kilograms possesses when it is travelling at a speed of one metre per second. We call this unit the 'joule'. It is the number of joules of electrical energy or chemical energy in gas that we use up, mostly by producing heat one way or another, which determines our electricity and gas bills. How big these are depend upon how rapidly our bits of domestic equipment eat up energy. All electrical equipment is rated by the number of joules it converts per second, so one can always work out how much it is costing us a second when the machine is switched on (an exceedingly unnerving thing to do, incidentally). The rate of energy consumption (strictly, conversion, since energy is always conserved) is what we call power, and the unit of power is the watt, equal to one joule per second. One bar of your electric fire, rated as one kilowatt, converts no less than 1,000 joules of electrical energy into heat every second. Your own body produces heat at about an eighth of this rate: eight people in a room equals one bar of an electric fire. With world energy resources at a premium, the joule, as the unit of energy, is a vital measure in the fields of politics and economics as well as physics.

When it comes to the transmission of energy by radiation we are naturally concerned with rate, rather than total amount. But, away from the source of radiation, in some region of space surrounding the transmitter, it is not the total rate of emission of radiation that concerns us, but the amount that reaches us. The useful measure here is the amount of energy per second which falls upon a square metre, and we call this the intensity, and measure it in units of watts per square metre. The intensity of radiation from the sun just outside the earth's atmosphere is $1 \cdot 3$ kilowatts per square metre ($1 \cdot 3$ KW m^{-2}). The intensity of a powerful laser beam can be a staggering 10^{10} KW m^{-2}. Intensity is the most useful measure, but it is not the only one. Sometimes we are concerned with the amount of energy in a given volume of the radiation, and so we use the concept of energy density, the number per cubic metre existing at any instant in the beam. It is easy to show that if we multiply the energy density by the velocity of the waves of which the radiation consists we get the intensity, so these quantities are intimately related.

While we are on the topic of waves a mention ought to be made of the transfer of momentum, as well as of energy. Anybody who has been hit by a wave on the sea-shore knows that waves carry momentum. All waves, whatever their nature, carry momentum as well as energy. This makes itself felt whenever they are reflected or absorbed, for in both cases there is a rate of change of momentum, and that means a force. We can use this fact to measure the rate of transmission of momentum in radiation by defining the force per unit area, i.e. pressure, exerted on an absorbing surface, and we speak of radiation pressure. The radiation pressure exerted by sunlight is minute, but in huge, massive stars the radiation may be so intense that the pressure it exerts on the electrons and protons of which the star is composed is the major prop against gravitational collapse.

The dynamical concept of momentum and energy can be applied to waves quite easily, but note how different the dynamic behaviour of waves is (Figure 27). Increasing the momentum and

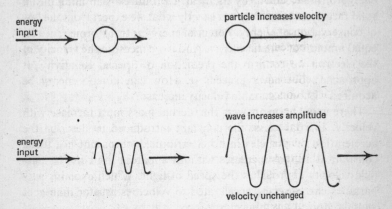

energy input → ○ →   particle increases velocity ○ →

energy input → 〜〜〜 →   wave increases amplitude 〜〜〜 →
velocity unchanged

Figure 27. Difference in dynamic behaviour of particles and waves.

kinetic energy of a particle always entails increasing its velocity. Speeding up is the particle's way of assimilating energy. The velocity of a wave, on the other hand, does not change. It is fixed by the properties of the matter through which the wave propagates. A wave assimilates energy without speeding up. Its amplitude increases, but its velocity remains constant.

But oddly enough particles start behaving dynamically rather like waves when their speed approaches the velocity of light. As we saw in the chapter on motion, strange effects make their appearance when particles reach high speeds. One of these is that no amount of force can accelerate a particle beyond c, the velocity of light in a vacuum. An electron in a linear accelerator is pushed along by a constant electric force, at first picking up speed rapidly, and then picking up speed more and more slowly as its velocity approaches c. Here is a terrible challenge to our concept of potential energy. Surely the electron's kinetic energy must increase by exactly the amount the electron has dropped its potential energy. Otherwise our law of conservation of energy absolutely fails. And yet how can it do so, if its velocity is limited by the velocity-of-light barrier?

The paradox becomes needle sharp when we measure the kinetic energy of the electron (by the heat it produces, slamming into a solid target) and find that it is exactly what we expect from the law of conservation of energy. Potential energy *is* transformed into an equal amount of kinetic energy. And if we measure the velocity of the electron we confirm the prediction of Special Relativity. It approaches, but never exceeds, c. How can kinetic energy be acquired without a suitable velocity increase?

There is only one answer: the inertial mass must increase with velocity. Inertial mass, the constant introduced to describe the acceleration of particles in an electric field, turns out not to be constant at all, but increases without apparent limit as the particle velocity approaches the speed of light. Which explains why particles can never be accelerated to velocities greater than c; it would require an infinitely large force.

This dependence of mass on velocity forces us to look at our concept of kinetic energy with new eyes. We never really understood in any case why kinetic energy was the peculiar quantity $\frac{1}{2}mv^2$. Why not mv^2 or $\frac{1}{3}mv^2$, or something else? Now perhaps this variation of mass with velocity will make that odd formula more acceptable. And so it turns out. All we do is to define kinetic energy in terms of the increase in mass rather than velocity. We define kinetic energy as the product of the increase in mass and the square of the velocity of light – in symbols $(m-m_o)c^2$, where m is the mass of the particle at any velocity, and m_o is the mass of the particle when it is at rest. The way in which the mass varies with velocity ensures that this formula yields $\frac{1}{2}mv^2$ at low speeds, so this definition is entirely compatible with, besides rendering intelligible, the older definition. We had the wrong concept of kinetic energy. We were mesmerized by speed because mass appeared to be a constant. Kinetic energy is not primarily to do with velocity. It is to do with increasing mass, and is, indeed, directly proportional to it. Energy of motion manifests itself directly as increase of mass. When we run we get heavier!

But there is more to come. If part of the mass of a moving body is energy, why not all of it? What is there to stop us conceiving of a quantity called the total energy, E, of a particle where $E = mc^2$, being composed of kinetic energy $(m - m_o)c^2$ plus a new sort of energy, intrinsic to the particle, equal to $m_o c^2$ which we call the rest-mass energy? The answer is, nothing at all. But if rest-mass energy is to have any real meaning it must be capable of being converted to other forms of energy in some interaction or other. After all, energy is a concept which is useful only in describing the way physical events in our universe occur. Are there events which transform rest-mass energy into other sorts of energy? Indeed there are (Figure 28). Electrons and positrons can annihilate one another, converting their rest-mass energies completely into electromagnetic energy in the form of γ-rays. In an atomic bomb, mass is explosively converted into kinetic energy. The fusion of two nuclei of deuterium (heavy hydrogen) to give a helium nucleus releases an enormous amount of energy directly related to the differences in rest-mass. Thus rest-mass energy is a real quantity. It is a type of potential or stored energy locked up in the particle, which under certain circumstances can be tapped. Energy and mass are essentially the same thing. Einstein's equation $E = mc^2$ expresses one of the most fundamental relationships of nature.

Fortunately the concept of momentum as mass multiplied by velocity does not have to be changed, provided the mass is the velocity-increased mass. The laws of conservation of energy and momentum remain valid. Nor do we have to change our concept of force, provided we stick to the definition that it is the rate of change of momentum with time, and not mass times acceleration. The essential advance arising out of the study of fast things is the expanding of our awareness of the intimate connection between mass and energy. And this raises several questions and opens up novel lines of thought. For example, if energy can be related to the mass of a particle is it possible to attribute mass to the potential energy of a field or to the energy of electromagnetic radiation? This will be pursued in a later chapter.

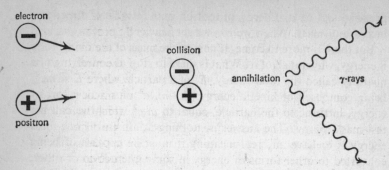

(a) matter totally converted into electromagnetic radiation: pair annihilation.

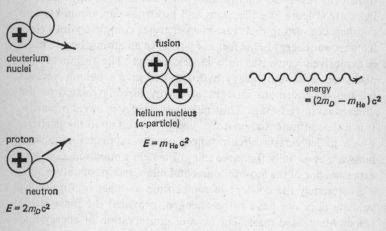

(b) fusion: Rest-mass of helium nucleus less than sum of the rest-masses of two deuterium nuclei. Difference available as energy.

Figure 28. $E = mc^2$.

We have culled the vital concepts of energy and momentum out of our study of simple gravitational and electromagnetic interactions. Let us close this chapter with a comment on some numerical coincidences. The ratio of the electrical attraction to gravitational attraction of a proton and an electron is of the order of 10^{40}, a huge dimensionless number. But a familiar one. Recall that at the end of chapter 3 we pointed out that the ratio of the radius of the universe to the Compton wavelength of the proton was also of the order of 10^{40}, and in chapter 4 we estimated the ratio of the age of the universe to the characteristic time of the nucleus to be again of the order of 10^{40}. Coincidence, or vitally significant? Whichever it is – and there was no doubt in the minds of many physicists, notably Eddington, that it was the latter – this number crops up again as a dimensionless quantity formed out of the following fundamental constants: G, the gravitational constant; m_p, the mass of the proton; $\hbar$, Planck's constant divided by 2π; and c, the velocity of light; viz $\hbar c/Gm_p^2$. Furthermore, the estimated number of particles in the universe is 10^{78} which is near its square ...

CHAPTER 7

FREEDOM

Lord Ronald . . . flung himself upon his horse
and rode madly off in all directions.

Stephen Leacock: *Gertrude the Governess*

THERE is always error in a practical measurement. That cannot be
avoided – it is just something that has to be taken quantitatively
into account and added to the result of the measurement, as a
measure of its accuracy. Our equipment is always too crude, and
our control over the conditions under which the experiment is
carried out, always inadequate. Improvements of technique, new
methods, fresh knowledge continually reduce errors and increase
accuracy. How far can this progress proceed? Can we *in principle*
extrapolate this process as far as we wish? Are we allowed to hold
the concept of quantities which can be determined with zero error?
Yes, says classical physics. Not without paying for it elsewhere says
quantum mechanics.

The trouble lies with the quantum of action, plus the fact that
we have to use one bit of the universe to measure another. If we
want to measure some property of a particle – its energy or
momentum, or its position in time and space, we have to allow the
particle to interact with our measuring system. Interaction means
exchange of energy–momentum. We therefore inevitably disturb
what we are trying to measure by the act of measurement. Obviously the disturbance must be kept as small as possible. No
problem in principle for classical physics – disturbances can be
made infinitesimally small, and energy, momentum, position and
time can all simultaneously be measured to arbitrary accuracy.
Not so for quantum theory.

Suppose we wish to measure the kinetic energy of an electron –
for simplicity, a slowly travelling one, so that we can forget about
special relativistic effects. The interaction of the electron with our

measuring system can be idealized by regarding the essence of our system to be another elementary particle – for instance, a photon. The photon, after interacting with the object, gives up its information about the object in some way that is rendered visible to us at the macroscopic level. To keep the disturbance low, the measuring particle, in bouncing off the object, must impart as little energy as possible. No problem enters here – the photon can be one with as low an energy as we like. But the photon carries one unit (h) of action – energy multiplied by time – and low energy means a long-period wave. The time of the measurement cannot therefore be ascertained more precisely than this wave period. Energy can be measured accurately but time, in the same measurement, cannot. If, on the other hand, the time between events involving the electron is to be measured accurately, photons with short-period waves (high frequency) must be used, with the consequence that the energy disturbance will be large. Energy and time are, in this sense, conjugate.

We now have a very strange state of affairs. The evolution of our concepts of energy and momentum has proceeded hitherto quite independently of the evolution of our concepts of space and time. Space–time was one thing and energy–momentum another. But now we find that because of the existence of the quantum of action the measurement of one of these entities materially affects the measurement of the other. They can no longer be considered as separate, independent, quantities: energy pairs with time, and momentum pairs with distance. Heisenberg summed the situation up in his uncertainty relations (Figure 29). In any physical interaction there is always a finite amount of action, which cannot be less than h (Planck's constant) associated with the uncertainties of energy and time, or momentum and position. The product of the uncertainties of energy and time cannot be less than h nor can the product of the uncertainties of momentum and distance be less than h. At best, these products are equal to h. Accuracy in energy has to be paid for with accuracy in time; accuracy in momentum costs accuracy in distance.

Nor can we avoid this mercenary relationship by using particles other than photons. Each particle carries its unit of action like the

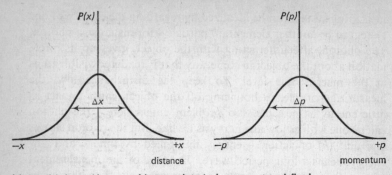

(a) particle located in space with uncertainty Δx has momentum defined with uncertainty Δp, such that $\Delta p \Delta x = h$.
($P(x)$ = probability of particle being at x; $P(p)$ = probability of particle having momentum p.)

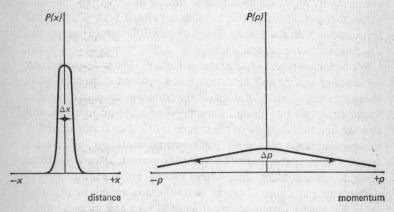

(b) particle more accurately located has more uncertain momentum.

Figure 29. The Uncertainty Principle.

photon, and the principle remains the same. Further, if we perform measurements of rotation we find a similar conjugacy of angular momentum and angular position.

One vital consequence of uncertainty is that the laws of physics have to be statistical in character. Quantum theory has to talk about expectation values rather than determined quantities, probabilities rather than certainties. Prediction is more uncertain, effect more diffusely related to cause, especially in situations involving single events. The classical orbit of an electron in an atom does not exist – there remains only a pattern of probability of finding the electron in a given region of space. The conservation of energy and momentum holds *on average*, but in individual cases energy may be created out of nothing or it may disappear without trace. Nevertheless, it is most probable that energy is conserved. The dogmatic absolutism of the classical law has to be replaced by the more realistic probabilistic statements of quantum mechanics. Though a strict determinism disappears, detailed models for interactions of all kinds can get generated by the quantum theory, pioneered by Schrödinger, Heisenberg, Pauli and Dirac, and these models have a quantitative precision which can be remarkable. Chance is no stranger to classical physics, as we shall find out in a later chapter. Whenever large numbers of particles are involved we are reduced to measuring average quantities and deviations from the mean, and so forth. It is quite impossible practically to know the position, momentum and energy of each particle at any given instant. Now we know we cannot, *even in principle*, determine precisely both position and momentum, and we have had to push our statistical ideas down to the level of the individual particle.

And all of this is necessary because elementary particles possess a quantum of freedom. Events for them are not strictly determined as they were for particles in the nineteenth-century classical physics. An element of choice appears to be theirs, to be used within certain prescribed limits. All is not allowed, but little is entirely forbidden. Idiosyncrasies are tolerated provided that, on average, the classical laws are obeyed. And the result of this freedom is a story of such peculiar effects, ideas, and consequences that one might be forgiven for preferring the comparatively everyday homeliness of science fiction.

Freedom is guarded jealously. Try confining an electron inside an atomic nucleus. After all, why should not electrons be a component of nuclei as well as protons and neutrons? Neutrons suffer radioactive decay and change into protons, and what comes out in the process is an electron. So an electron trapped by a proton to form a neutron seems a reasonable idea. The electron will have none of it, however. It refuses to cooperate. It declines to be confined. It has its freedom to think about. An electron confined to a volume of space of nuclear dimensions must, of necessity, have wavelengths associated with it at least as short as the diameter of the nucleus. If the waves were much longer it would mean that the electron spends most of its time outside of the nucleus and that would not do. But short wavelengths imply a restriction in space, and that must be balanced by an expansion of momentum in order to retain its fundamental quantum of action, h. The electron would then have so much kinetic energy it would burst out of its nuclear cage. The imprisonment cannot be done. Electrons cannot exist inside the nucleus in any stable state, unless some tremendous force is brought to bear to overcome their push towards freedom. Such a force as the immense pressure inside a star collapsing under its own gravity can squash electrons into nuclei to form a body composed entirely of neutrons – the neutron star. And that is a graphic measure of how strong the freedom urge of the electron is. It needs a body of the size of a star to sit on it.

Every time we try and constrain the electron in some way, either by forcing it into a confined space, or by directing it through slits, it insists on its freedom of action and displays it in a characteristic way. But the electron is not restricted to passive demonstrations. It can exploit its freedom by violating the laws of conservation of energy and momentum.

The source of all electromagnetic radiation is the acceleration or deceleration of charged particles, mostly electrons. Before there can be light there must be electrons with energy to spare. Since the energy in light comes in packets called photons, the basic light-producing process must be the emission of a photon by an electron. Of course, once the photon appears, it rushes off at the speed of light, carrying with it an amount of energy characteristic of its frequency and an amount of momentum characteristic of its wavelength. The electron, reduced in energy, recoils like a gun which has just fired a bullet. Provided the electron has enough kinetic and potential energy to begin with, and provided that it can cope with the recoil, as it can in an atom by pushing against the nucleus, the emission carries the seal of approval of the laws of conservation of energy and momentum, and it is allowed to take place. If the electron did not have enough energy, or if it cannot cope with the recoil, the emission of a photon just does not happen. Such is the situation for a free electron: it is forbidden to emit a photon by the conservation laws.

That might be acceptable to an old-fashioned particle, but to a modern particle like an electron it is an intolerable constraint on its freedom. One can imagine the revolutionary declaration . . . 'It is the inalienable right of every free electron to emit or absorb photons . . . ', and slogans like 'Down with the conservation laws!' Nature finds herself with a sizable confrontation on her hands and deals with it brilliantly so that everyone wins. An electron is allowed to emit any photon it pleases (an enormous concession), provided it absorbs it within a time allowed by the uncertainty principle (sighs of relief from the conservationists). The mainstay of this system is the quantum of action associated with the electron–photon interaction: the time component of the quantum of action is the time between emission and absorption, the energy component is the photon energy. Thus a high-energy photon has to be absorbed very quickly after being emitted, but for low-energy photons the time between emission and absorption can be quite long. Under no circumstances can such photons escape, so overall there is no energy–momentum loss, and, on average, the conservation laws are satisfied. Only when energy and momentum are conserved can a real photon appear.

This remarkable activity in which an electron emits and absorbs a photon is called a *virtual process*, and the photons, which have an evanescent existence, are known as virtual photons (Figure 30).

Actually, virtual processes are somewhat analogous to that familiar human activity, borrowing. You have insufficient money to achieve some goal. The law of the conservation of value forbids it. Of course this does not stop you. You go to your bank manager and borrow £X, promising to pay it back in t weeks time. Within certain limits determined by your credit-worthiness, glibness, connections, etc., etc., the quicker you elect to pay back the more you can borrow. Roughly, the product of X and t is some constant (but

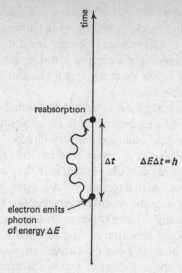

Figure 30. A virtual process.

not a universal one). With £X you get your goal but t weeks later you have to raise £X. For the analogy to be close, you would have to sell whatever it was you acquired for £X, and you end up exactly where you started, not even paying interest. In society, borrowing allows an extra freedom of action (to make virtual purchases). In nature, the quantum bank extends credit facilities to all elementary particles as of right, so that they may indulge their peculiar whims.

What virtual processes do photons indulge in? When γ-rays pass through matter, they can convert themselves into an electron–positron pair, if the photon energy of the γ-ray is enough to provide at least the rest-mass energy of the pair. In vacuum, the conservation laws forbid pair-production even for high-energy γ-rays. Photons of visible light have nowhere near enough energy to produce pairs anywhere. Yet a photon of whatever energy you like can convert itself into an electron and positron provided that they recombine rapidly to give the original photon (Figure 31). The transient electron and positron are known as a *virtual pair*.

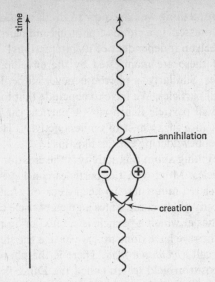

Figure 31. Virtual electron–positron pair production by photon.

Its existence has to be very brief indeed – something like 10^{-21} of a second – but as the photon continually indulges in pair-production it will be continually surrounded by virtual pairs.

In the same way, electrons surround themselves with a cloud of virtual photons. Indeed a further esoteric process may be envisaged as a consequence of the quantum of action. It is that virtual photons may surround themselves with virtual pairs! So an electron moves about in the centre of a cloud of photons *and* electron–positron pairs. Moreover, the electrons, during their transient existence, will be repelled electrostatically by their parent electron, whereas the positrons will be attracted. There will be therefore a shift of the positron cloud towards the electron and a shift of the electron cloud away. The electron has electrically polarized the vacuum! It has the effect of smearing the negative charge of the electron over a radius of about the Compton wavelength, 10^{-12}m.

There is an extremely serious consequence of all these virtual processes. We can no longer regard the electron as a single thing.

It consists of something we have now to think of as a 'bare' electron which is 'dressed' in a cloud of photons and pairs. We cannot think of an electron independent of its interactions! The strength and nature of these are manifested by the enveloping cloud of virtual particles. Similarly, a photon is never alone, it too has its cloud of virtual particles. We have to conclude that the concept of a free, individual particle is a myth. A particle carries its interactions with it in a fuzzy mist of endless activity. How must we now conceive of bewildering entities like this?

The essential thing about this activity is the creation and annihilation of particles. Moreover, the particles can have all possible energies and related momenta. To keep track of what is going on we have to specify all dynamic states a given particle can have and describe the rates at which a particle is created or annihilated in a given state. When we have done that we have a theoretical model, something we call a *quantum field*. There is the photon field and the electron–positron field (often called the Dirac field). Each is conceived to be all-pervasive and to manifest itself by the creation and annihilation of its particles. The virtual production of particle and anti-particle means that a real particle is tied physically to the field and can never be considered as a separate entity.

In the classical fields of gravitation and electromagnetism a region of space is conceived to have special properties which influence the motion of test-bodies. We speak of the force of a body experiences, and of its potential energy, at every point in the field. We can identify the source of the field as a massive body or an electrically charged one. Quantum fields are different. There is no source, they are present everywhere. They manifest themselves in the creation and annihilation of elementary particles, both real ones and virtual ones, according to the rules of quantum mechanics. A particle is to be viewed theoretically as an excitation of the field, and in no way independent of it.

Perhaps the oddest feature of quantum fields is their total inability to keep still. Like energetic children they are always jiggling. Serenity, calm, restful inactivity, are not possibilities. Even when no real particle is around, the field is gently, but persistently, bubbling with activity. Particles are continually being created and almost immediately destroyed, enjoying only a transient existence in a never-ending effervescence. The barely credible fact of the matter is that the field indulges in virtual processes quite spontaneously! Everywhere in the universe the electromagnetic field is busy creating ghostly photons out of nothing and just as busily annihilating them. Everywhere the Dirac field is demoniacally creating and annihilating electron–positron pairs (Figure 32). Nowhere is free of the activity. The vacuum is a very active substance.

It would be hard to overestimate the significance of this extraordinary activity. Some of its consequences are very comforting, but some are plainly horrific. One comforting feature is that the vacuum fluctuations of the photon field jiggle energy-rich electrons in atoms, and induce them to emit light 'spontaneously'. Otherwise we would live in a pretty dim and cold world. On the other hand, if we want to produce coherent light in a laser we have to fight against the random, chaotic emission induced by these fluctuations, so they can be a nuisance. Either way, they are very real things.

But the real horror comes when we appreciate the quantitative significance of vacuum fluctuations. Because, on average, each fluctuation amounts to having half a quantum of energy in every possible dynamic state – not a full quantum, because annihilation follows rapidly on creation, but not zero either. For every conceivable wavelength of electromagnetic radiation, there is half a quantum of energy. If we start off with some real photons of a given wavelength we can absorb them one by one until no real photons are left, but we cannot absorb and eliminate the half a photon associated with the fluctuation. That energy we call the *zero-point energy* for that wavelength. It cannot be touched. We cannot tap it to provide a source of power. It must always be there. The horror is that if we add up the zero-point energies of all the wavelengths we get an infinite amount! Very embarrassing. Every quantum field that operates in a vacuum has an infinite amount of energy in its fluctuations, before we even begin to add energy in the form of real particles. That is how it is, like it or not! We just have to accept it and tread very carefully when we are making

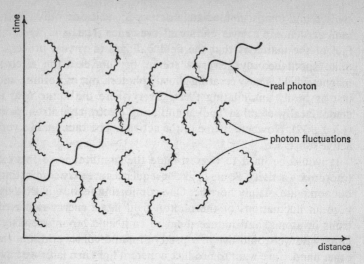

(a) the photon field.

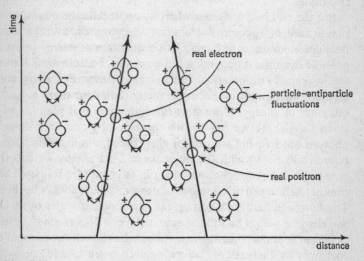

(b) the Dirac field.

Figure 32. Quantum fields.

theories. We will come across vacuum fluctuations again in the next chapter. Let us move on to explore other facets.

Another consequence of thinking in terms of quanta is that we must invent a completely new picture of how electrons interact with one another. The mutual repulsion of two electrons (like charges repel) is, in classical terms, the response of the electrons to each other's electric field – instantaneous action at a distance. In quantum-field physics this effect has to be something to do with the annihilation and creation of particles, which seems a far cry from the classical description! Nevertheless, that is the way it looks. What happens is that the electrons exchange virtual photons of a special kind, and the exchange continuously knocks them further and further apart. These special photons are the quanta associated with electromagnetic waves, which have their oscillating electric field pointing along the direction of propagation. Such waves are said to be longitudinally polarized. Ordinary electromagnetic waves have their electric fields at right angles to the direction of propagation and are said to be transversely polarized. The curious thing is that longitudinally polarized waves cannot exist according to classical theory. Photons of this type of radiation owe their existence to the quantum of action, and they can exist only as virtual photons. But they are exactly the particles required to explain the instantaneous electrostatic interaction between the two charged particles (Figure 33). Such an interaction is the consequence of a continual exchange of virtual photons.

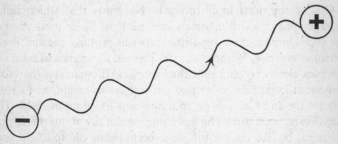

Figure 33. Instantaneous electrostatic attraction due to exchange of virtual photons.

This idea of exchange is used quite generally in quantum physics to describe interactions. We have mentioned the photon field but there are others whose virtual particles give rise to repulsions and attractions.

One such field provides an explanation of superconductivity. If a metal at very low temperatures loses all resistance to electric currents we call it a superconductor. Oddly enough we can explain this effect by an attraction between electrons, which comes about through their interaction with the mechanical vibrations of the metal atoms. The energy of vibrational waves comes in packets which we call phonons. A phonon is to sound what a photon is to light. So here we have another quantum field, the phonon field. This is somewhat different from the other fields we have mentioned in that it does not live in a vacuum but only in macroscopic matter. Nevertheless, the phonon is a fully-fledged quantum particle which plays a fundamental role, particularly in the determination of the properties of solids. The attraction between electrons in a superconductor comes about through the exchange of virtual phonons. There is a certain irony about this, because the emission and absorption of real phonons by electrons is the basic cause of electrical resistance in metals. At low temperatures electrons can get around this restriction on their free motion by, as it were, gobbling up phonons as soon as they are emitted so that they keep what motion they have acquired to themselves. Electric currents in a superconducting electrical circuit, once started, can then continue to flow without hindrance.

One of the most basic forces in Nature is that which holds together protons and neutrons in a nucleus. Since it overcomes the formidable electric repulsion between protons, we call it *the strong interaction*. Without it, nuclei would fly apart and no more complex element would exist than simple hydrogen. It is therefore somewhat important to try and understand its origin, and naturally we try and pin it down to a new sort of quantum field. The question is, what sort? The first thing we can say about this is that it cannot be like the photon field, because the photon field produces long-range forces, and the strong interaction has to be a short-range affair, effective only over distances of the order of

10^{-15}m. The basic reason why electric forces reach out so far is bound up with the fact that the photon has no inertial mass whatsoever. It is worth while trying to understand why this is so.

Suppose we have two electrons a kilometre apart and ask how the exchange of virtual photons works. A photon emitted by one travels at the speed of light to the other and is absorbed. The time taken is quite long, as times in the quantum world go, and is a considerable component of the amount of action allowed for a virtual process. The energy of the photon has therefore to be pretty small, because energy multiplied by time cannot be much different from h, Planck's constant. This is no problem for the photon field because the energy of a photon can be as small as it likes. There is no limit because the rest-mass energy of the photon is zero. So the two electrons can be as far apart as they like and still indulge in this quantum game of exchange, because there are always photons of sufficiently small energy to play with. A quantum whose rest-mass is not zero cannot have less energy than its rest-mass energy, so it cannot possibly be exchanged in a virtual process, if the time between emission and absorption gets too large. The particles doing the exchange must not be too far apart.

We can now see that the things which are involved in the strong interaction must have a mass, because that will automatically reduce the range at which exchange can occur. Since we know the range we can calculate the mass, and it turns out to be some 200 times bigger than the mass of the electron. Real particles of this mass are known to exist – they are the π-mesons. Moreover they come as positively charged, negatively charged, and neutral particles, and this helps to understand how they can interact equally well with positive protons and neutral neutrons. Happily then, we can associate the strong nuclear force with a meson field. Protons and neutrons play the exchange game with virtual mesons (Figure 34). A nice prediction, and a neat concept. Unfortunately, there are two other types of meson – the κ-meson and the η-meson, both about 1,000 times heavier than the electron – and they both interact strongly with nucleons. Are these associated with an even shorter-range force? Or are they in some way excited states of the π-mesons?

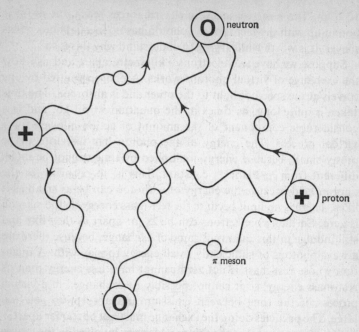

Figure 34. Binding of protons and neutrons in the nucleus by the virtual exchange of π-mesons.

And what about all those heavy particles created in high-energy collisions – the Λ-hyperon, Σ-hyperon, Ξ-hyperon and the Ω-hyperon? They look like heavier versions of the proton and neutron and so we can let them use the meson field to indulge in their strong interactions. But the meson field cannot be pushed too far. There exists a whole batch of processes involving hyperons and, indeed, mesons themselves, which cannot be explained without invoking yet another quantum field. The fact is that none of the hyperons are stable and nor are the mesons. Even the neutron, once outside of the nucleus, is unstable. Hyperons decay to other hyperons and mesons in times of the order of 10^{-10}s. Neutrons decay into protons which are quite stable. The decay products from hyperons, mesons and neutrons include muons, electrons, neutrinos and photons. What makes them decay? Whatever the force is, it has to be considerably weaker than the strong interaction, and it turns out to be even weaker than electromagnetic forces by a factor of some 10^{11}. We call it the *weak interaction*. If a quantum field is at the base of this interaction, it must have a particle associated with it whose mass is even greater than the proton mass. No particle with the right properties has yet been discovered, so the weak interaction is something of a mystery.

The only force we have not mentioned as a candidate for a quantum field interpretation is the weakest of all, gravity. But the effects of the tiny quantum of action are so small that only a respectable urge towards completeness moves the theoreticians to worry about quantizing the gravitational field. Like the electromagnetic field gravitation is a long-range affair, so its quantum particle cannot have any mass. We call it the *graviton*, and nobody has observed such a thing. But we like to think of it being around, nevertheless. The strong and weak interactions provide a rather more immediate challenge in the quantum world, and to these we must return.

We cannot investigate these new interactions in the same way as we investigated the electromagnetic and gravitational forces, for, being short-ranged, they have no classical macroscopic effects whatever. All that can be done is to observe the patterns of creation and decay, and observe which occur and which are forbidden. The important pattern of familiar interactions is summarized in the conservation laws for energy, linear and angular momentum. Fortunately these laws are found to hold for the new interactions. But familiar interactions can give no guide to the sort of particles which can occur in particular creation and decay processes. We have to proceed empirically and deduce additional laws, expressing quantities which do not change.

One interesting thing about conservation laws is that they are associated with symmetry. Thus the conservation of energy is associated with the fact that the interaction laws remain the same, whatever zero of time is taken. The laws, we say in the jargon, are invariant under time-translation. The conservation of linear momentum comes about because of invariance under space-translation, and the conservation of angular momentum arises because the interaction laws do not change when the coordinate system is rotated to a new position. Another example is the conservation of charge, which is intimately associated with the fact that electromagnetic laws work independently of the zero of electrical potential. As far as we can tell, *all* interactions observe these symmetries. What symmetries are implied by the nuclear conservation laws we are beginning to manipulate. The omega minus (Ω^-) hyperon was predicted by Gell-Mann in his so-called eight-fold way. But whatever symmetries they are they are not shared with the weak interaction – the decay of particles does not obey the rules which govern strong interactions.

The weak interaction, in fact, is more famous for its lack, rather than its abundance, of symmetry. Another sort of symmetry is that provided by mirror images (Figure 35). In a mirror, right hands become left hands, spins go the other way. If a property of a particle changes sign under such a reflection, we say the property has an odd parity, whereas if it does not change sign we say it has even parity. The strong and electromagnetic interactions are interdependent of space reversal and so parity is a property which is conserved in all such processes. It was a shock to physics that this seemingly basic symmetry was violated by weak interactions – parity is not conserved. One might also expect that interactions would go equally well if all positive charges were replaced by negative charge and vice versa – displaying symmetry under charge reversal. Not so the weak interaction.

We know that Nature sometimes prefers, in a given system, to be left-handed, or right-handed. Most of us are right-handed. The point is that it is not forbidden to be left-handed. There is nothing deeply embedded in physics that prevents a plant cocking a green snook at its spiralling fellows and climbing anti-clockwise towards the sun. It is only the weak interaction which appears to impose this lack of parity symmetry.

And not satisfied with that it goes on to do the same thing with electric charge. This behaviour is very hard to understand, and to formulate a theory which is consistent with special relativity and electrodynamics without these basic symmetries is extremely difficult. One way out of the dilemma is to enter boldly into pure

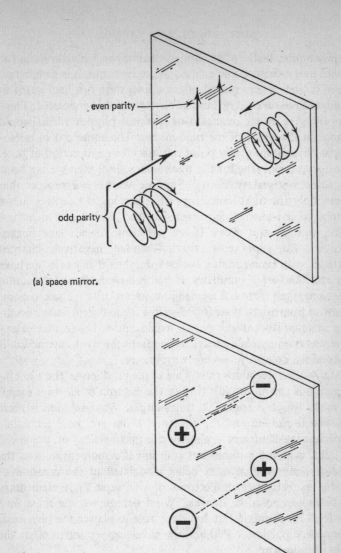

even parity

odd parity

(a) space mirror.

(b) charge mirror.

Figure 35. Mirror images. The strong and electromagnetic interactions do not change when we enter either the space mirror world or the charge mirror world. The weak interaction does.

fantasy and consider what happens if time ran backwards. In the case of every microscopic process we know about, it is a plain fact that if a ciné-camera record of the events were run backwards to simulate time-reversal no laws of physics would be violated. There is nothing in either quantum or classical physics which would change in the world of the time-mirror. The equations of microscopic physics in no way point out a preferred direction of time. But if that is not true for the weak interaction, then we can hope for some vestigial symmetry. Suppose we have a process that works only for right-handed, positively charged particles. If we change to left-handed, positively-charged particles it does not work, nor does it work if we substitute negatively charged particles. But suppose we try left-handed, negatively charged particles with time running backwards, then if it works we have discovered a sort of symmetry. If that is how the weak interaction functions, then there is a tremendous implication – a microscopic arrow of time exists. If we found such a triple-mirror image occurring in some distant galaxy, we would know that in that galaxy time was running backwards! But perhaps the weak interaction is just plainly, doggedly, lacking in symmetry.

Many other questions exist. One of the products of the weak interaction is the muon, identical to the electron in all ways except its mass, which is some 200 times larger. What relation is there between the muon and the electron? What are those incredible particles, the neutrinos – ghostly things, travelling at the speed of light, carrying nothing but spin angular momentum and the memory, in some way, of being associated in the creation of muons as distinct from electrons or vice versa? Are elementary particles composed of quarks? What determines the mass of a particle? Does gravitation have any role to play in the physics of elementary particles? Perhaps we have not yet learnt what the right questions are.

CHAPTER 8

MASS

Consider anything, only don't cry!
Lewis Carroll: *Through the Looking Glass*

LET us return for a little while to pre-quantal innocence, and think about the electron as a charged billiard ball. In chapter 6 we pointed out that the concept of inertial mass was needed to explain why electrons and singly-charged ions in an electric field are accelerated by different amounts and how inertial mass is involved in definitions of kinetic energy and momentum. But what is inertial mass? We know it is a measure of inertia, the property a body has to resist any change in its motion; but that is saying nothing but a tautology, another way of putting it, but nothing new. Let us ask, instead, can we relate inertial mass to some other property of the particle? If we can, then things will be simpler than they were before. Two properties, one of them familiar inertia, will be seen to be the same entity.

It is a fascinating fact that classical electromagnetism could do just that! Think of the billiard-ball electron moving along with constant velocity. It is a moving charge and therefore, surrounding it throughout all space, is a magnetic field, as well as the usual electrostatic field.

Now the magnetic field surrounding the moving electron possesses potential energy, and we can add up this potential energy throughout the field, to get the total amount of energy residing in the space, stretching from the surface of the billiard-ball electron out to infinity. The answer we get is larger the smaller the radius we assume for the billiard ball, because the field near the ball gets more and more concentrated the smaller we make the ball. But this energy is entirely due to the motion of the electron – it would be zero if the electron were not moving, because magnetic fields appear only when charges move. It has got to be regarded, therefore, as part of the kinetic energy of the moving electron. And indeed this energy increases as the square of the velocity, at low velocities, just as kinetic energy does. But since kinetic energy is intimately related to inertial mass we are forced to conclude that part of the total inertial mass of the electron resides in the space all round the electron. Stopping the electron means destroying its magnetic field, so its inertia has to be, in part at least, associated with that field. This conclusion remains valid today ... But if part, why not all? It turns out that if we assign a radius of about 10^{-15}m to the electron then, indeed, all of the inertia is explained by the magnetic field. If that is so for an electron in motion, what can we say about the rest-mass energy that special relativity tells us a stationary electron has? Given that the electron is a negatively-charged, utopium billiard ball, then the answer is simple. The rest-mass energy is just the electrostatic energy arising out of one bit of the charge repelling all the other bits. The electron would like to explode, but something holds it together and there it sits, full of pent up energy. The total energy of a moving electron, rest-mass plus kinetic, is therefore nothing but the total energy of its own electromagnetic field.

This is a tremendous simplification. All inertia, in other words all the phenomena of mechanics, now come within the realm of electromagnetism. Bouncing billiard balls, oscillating springs, gyroscopes – all of these owe their behaviour solely to the fact that they are composed of elementary electric charges. Considering that we had to introduce the idea of inertial mass to account for the motion of charges (chapter 6) we perhaps ought not to be too surprised. But something truly exciting follows on from this, for have we not identified inertial mass with gravitational mass? If the 'charge' which produces gravitational fields, the quantity we call gravitational mass, is identical to inertial mass then gravity itself is an electromagnetic phenomenon! Electromagnetic fields attract electromagnetic fields. But such a property is not described by our equations of electromagnetism. We need a unified field.

Attractive though this model is, it cannot be accepted as it stands. Even within its own classical-relativistic terms of reference – that is, without bringing the quantum into it – there is some unease.

Special relativity, with which electromagnetic theory is entirely consistent, demands that $E = mc^2$ holds in a perfectly general way. But in order to satisfy this relation the model has to assume that the electron has a particular charge distribution, and, in fact, not a very simple one. Moreover what holds the repelling bits of negative charge together? Another force, another field? Certainly not an electromagnetic one. And then, of course, there is the quantum of action and the concomitant uncertainty principle, which does not allow us to contemplate in any meaningful way an electron radius of around 10^{-15}m without at the same time accepting that the corresponding uncertainty in momentum, and therefore energy, has to be very large.

Nevertheless something can be salvaged. It remains true that at least part of the inertial mass has an electromagnetic origin. How big a part depends upon the structure of the particle, and this has to be deduced ultimately from quantum field theory. The problem arises because we try to push classical concepts down to distances like 10^{-15}m. The question is: How big really is an electron? It is no use looking to see how big it is using ordinary light, because the wavelength of ordinary light is too long. We have to use γ-rays. But we hit a basic snag here because each γ-ray photon packs a considerable punch, and any electron which reflects one into our hypothetical γ-ray microscope gets such a knock it recoils furiously. So we never 'see' a stationary electron, we always see one that is moving away from us. Any radiation that comes from an object moving away from us has its pitch – its frequency, that is – lowered, and this Doppler Shift, as it is called, means that the γ-rays we 'see' will have a longer wavelength than the ones we used to illuminate the electron. The shorter the wavelength we use, the bigger the recoil and the greater is the shift towards longer wavelengths of the reflected rays. It turns out that the smallest wavelength that ever gets into our microscope is about 10^{-12}m. It is called the Compton wavelength of the electron and it represents the absolute limit of resolution. If the electron is smaller than this we can never find out how small using photons. But it does give us an upper limit, and if we take the Compton wavelength to be the radius of the electron, we at least stay more or less within the validity of our classical theory. Working out the magnetic-field energy of a moving electron, we come to the conclusion that the fraction of inertial mass which can be confidently taken to be of electromagnetic origin is $\frac{1}{137}$ (Figure 36). Interestingly enough this fraction turns up in the theory of atomic spectra, and there it is known as the fine-structure constant.

At least some part of inertia is of electromagnetic origin. What about the rest? A clue is provided by the motion of electrons in crystals. Electrons move around in crystals as if they had a mass

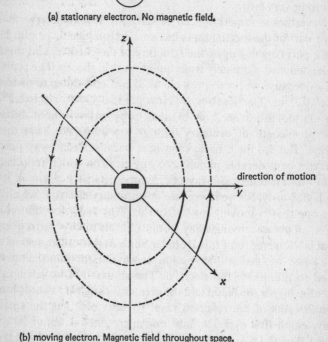

(a) stationary electron. No magnetic field.

(b) moving electron. Magnetic field throughout space.

Figure 36. Electromagnetic inertia. A fraction $\frac{1}{137}$ of the kinetic energy of a moving electron resides in the surrounding magnetic field. A fraction $\frac{1}{137}$ of the inertial mass of the electron is therefore in the form of electromagnetic energy.

entirely different from their ordinary mass. Not only can this 'effective' mass differ from the free mass by factors of ten or even a hundred, either way, it may assume different values in different directions, and, to crown it all, it may be negative! All of this is well-understood and in no way mysterious. These oddities come about because the electron is continually interacting with countless numbers of atoms, and so it is hardly surprising that its motion is radically different from what it would be in a vacuum. No one expects a billiard ball, attached by a spring to something large and solid, to exhibit the same response to a push as a free one. An electron in a crystal is attached to all the atoms by invisible electromagnetic springs, so of course its behaviour is bound to be different to that when free. The point is that the difference is not one of quality, or even of degree, but of environment. Even in a vacuum an electron is never free of invisible electromagnetic springs.

Quantum-field theory tells us what those invisible springs are. First of all, there is a spring between the electron and the Dirac, electron–positron, field. Bubbling around the electron is a seething mass of virtual electron–positron pairs. The spring is essentially an electrostatic one, it repels virtual electrons and attracts virtual positrons. We have the phenomenon of vacuum polarization, which we mentioned in chapter 7. Virtual positrons cluster close to the real electron and effectively neutralize its charge. The virtual electrons, repelled by the real electron, spread out around it in a sphere of radius equal to the Compton wavelength. So, in effect, the size of the electron, as far as electric charge is concerned, is indeed the Compton wavelength. So our conclusion that only $\frac{1}{137}$ of the inertial mass is electromagnetic is confirmed.

But perhaps that is wrong, because we have not considered the magnetic and electrical energy associated with the spin of the electron, and also there is another spring, that connecting the electron with the vacuum fluctuations of the photon field. When all of these factors are carefully taken into account the result is very simple: the spin energy is negative and the jiggling energy is positive and they effectively cancel one another out. Our conclusion stands.

The search for the origin of inertia in electromagnetism has not been wildly successful. We have accounted for only $\frac{1}{137}$ of it. Where can we look for the rest? There are three choices, the strong interaction, the weak interaction, and gravitation. The strong interaction certainly modifies the mass of atomic nuclei and may indeed contribute to the mass of isolated hadrons (baryons and mesons). But it is not easy to see how these interactions can determine the mass of leptons like the electron and muon. On the other hand gravitation affects all particles, so perhaps a promising line is to look for the basic source of inertia there. It is rather an odd idea that gravitation may in some way be responsible for inertial mass, since mass is certainly responsible for gravitation. But the intimate connection between mass and gravity is a strong indication that the most basic source of inertial mass is ultimately to be found in the most evident of all interactions, gravitation.

Even if that is wrong it is worth exploring. Not only will we emerge from such an exploration with our concept of space and time profoundly modified, but also we will find it difficult to avoid being exhilarated by the idea of that oddest of all astrophysical things – the black hole.

It all stems from what Einstein has called the principle of equivalence. Gravitational mass is introduced as a concept in much the same way as electric charge is in electrostatics. It provides a measure of the strength of gravitational field surrounding a body or, in basic operational words, the magnitude of accelerations induced in surrounding free bodies. Inertial mass, on the other hand, is, at base, a concept to describe the acceleration of charged bodies in an electromagnetic field (chapter 6). In a purely gravitational situation (individual particles interacting only via gravitation, thereby ruling out rigid bodies) the concept of inertial mass is not required. Nevertheless, in spite of their entirely different origin, it is found experimentally that, to a high degree of accuracy, the gravitational mass of a body is equal to the inertial mass. If it is assumed that they are identical, then it follows that the behaviour of things in a laboratory situated in a uniform gravitational field ought to be identical to the behaviour of things in a gravitation-free laboratory which is being uniformly accelerated. In a local region of space, gravitation and an appropriate acceleration are entirely equivalent.

Imagine experiments in a stationary lift. To a good approximation the gravitational field is uniform. We feel the pressure of the floor on our feet, apples drop vertically downward, projectiles follow parabolic paths. Now switch off gravity. We float; apples, once released, remain where they are, projectiles travel in straight lines. Obviously there exists a totally different physical situation. Now keep gravity switched off, but allow the lift to be pulled upwards with uniform acceleration. Once more we feel the pressure of the floor, apples drop vertically downward, projectiles follow parabolic paths. Is the lift accelerating really, or has gravity been switched on again? The principle of equivalence asserts that no experiment carried out entirely within the lift can possibly distinguish between the alternatives.

Obviously the principle has practical application, when it comes to manufacturing an artificial gravity aboard an interplanetary spaceship. But it is equally obvious that no simple acceleration can substitute for the Earth's gravity – the acceleration in New Zealand would have to be nearly opposite to that in England. Clearly the non-uniform gravitational field surrounding Earth is a real entity, not to be eliminated by theoretical accelerations. So what is the point of the equivalence?

The point is that by *thinking* of the gravitational field in terms of an acceleration, we can deduce that properties normally associated with motion are also properties of gravity. Thus we know that a ray of light which enters our *accelerating* life horizontally will leave the lift at a point in the opposite wall slightly below the level at which it entered, because during the time the light travels from wall to wall the lift moves upwards (Figure 37). The principle of equivalence tells us that the same behaviour will be observed in a stationary lift in a gravitational field. Gravity bends light.

The equivalence of pure gravitation and an appropriate acceleration leads to bizarre predictions. Imagine a space laboratory orbiting close to a large mass (for comfort, a cooled star) where the gravitational field is intense. They carry out experiments in physics, which we can observe from our relatively gravitation-free

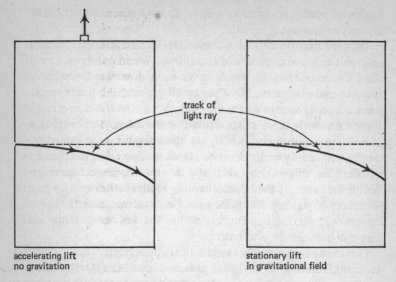

Figure 37. Equivalence of acceleration and gravitation.

observatory. The principle of equivalence explicitly states that the physical situation would be the same if the gravitational field did not exist and the space laboratory were subject to an equivalent acceleration. If indeed the laboratory were continually subjected to such an acceleration, which at every point was equivalent to the gravitational field at that point, it would have a well-defined velocity by the time it reached our gravitation-free observatory, and we would observe all the peculiar effects that special relativity predicts. We would see their clocks running slow, lengths contracted and masses increased. Such an acceleration therefore implies all sorts of special relativistic effects. If indeed it is to be physically equivalent to the gravitational field, then we should observe exactly those special relativistic phenomena in the space laboratory, when it is not being accelerated but is actually situated in the gravitational field. Thus we should find that radial lengths, which are in the direction of the equivalent motion, are shortened, but transverse lengths remain unaffected; clocks run slow and masses are larger. If we observe the passage of a beam of light in the far-off laboratory, we should find that its speed is apparently reduced from what it should be. The size of these effects depends upon the equivalent relative velocity which is related unambiguously to the gravitational potential at the field point.

Predictions along the above lines, insofar as they have been tested in the comparatively weak gravitational fields in the solar system, have been confirmed.

There is, however, a more startling conclusion. There exists a limit to the gravitational potential, a critical value, at which radial lengths shrink to zero, clocks stop, and the velocity of light travelling radially equals zero. Immensely dense stars which achieve this critical condition become invisible, for light cannot escape from its surface. Such entities are known as black holes. Though invisible, their gravitational field makes them detectable in principle, but so far nobody has 'seen' one.

The possibility of this critical condition applying to all objects, whatever their mass–energy, can be assessed as follows. Considering a body as having its mass concentrated at a point allows us to define a radius at which the gravitational potential reaches the critical condition. This value, which depends upon the mass, $(2GM/c^2)$ is known as the gravitational radius of the body. The gravitational radius of the Earth is only 0·4 cm, of the sun, 1·4 km. Since each value is less than the actual radius of the body, the critical condition cannot occur. But what about elementary particles? The gravitational radius for the electron is about 10^{-57}m, for the proton about 10^{-54}m. Compared with the classical electromagnetic radii, these lengths are infinitesimally tiny. The possibility that gravitation plays a role in determining the structure of elementary particles cannot be ruled out; but our imaginations would have to be fortified to embrace the description of the electron as a negatively charged black hole.

At the other extreme of distance the concept of a critical gravitational potential leads to the view that the universe is finite. Imagine a sphere expanding around us to include more and more galaxies. Eventually the total mass enclosed by the sphere will be so large that light cannot escape. The radius of the sphere at that point defines the radius of the universe. It is computed to be of the order of 10^9 light-years, corresponding to a total mass of about 10^{51} Kg, or 10^{78} electrons and protons. This is a considerable prediction, but it does not end there. The universe is observed to be expanding, at least as far out as we have probed. If so it implies that the gravitational constant, the total mass, or the velocity of light, singly or in combination, vary with time in order to keep the actual radius equal to the gravitational radius. What is thrown up here is the whole possibility of fundamental constants of nature varying with time!

All of these remarkable effects, from the bending of light to the determination of the dimensions of the universe, follow from the equivalence of gravity and acceleration. But, on the way, we rode in a somewhat cavalier fashion over some delicate ground, and we really ought to note the damage and the repair-work, since this is equally remarkable.

Perhaps the most serious casualty is the velocity of light, which is no longer a constant, but dependent upon gravitational potential. This may be acceptable in itself, were it not for the invocation of a privileged observer, situated in a gravitation-free environment, who could observe this shift in velocity. No such observer can possibly exist. All real observers are exposed to the gravitational fields of their planets and also to the gravitational field of the rest of the universe. There is no escape from gravity. It is universal. The invention of a gravitation-free observer smacks strongly of metaphysical ideas like absolute space and absolute time. It offends basic ideas of operationality and relativity. The mention of space evokes a further criticism. Must we not use the tracks of photons to define straight lines in space, and hence the geometry to be used? If we say that light is bent in a gravitational field, it means that we are assuming that space is flat, that straight lines can somehow be determined without recourse to light, that the geometry is Euclidean.

All of this damage is eliminated in Einstein's theory of general relativity. The basic principle is that the laws of physics should be the same for accelerated observers and hence, by the principle of equivalence, for all observers in whatever gravitational field they happen to be in. In particular, the velocity of light remains a constant, and light travels in straight lines, but space is no longer Euclidean. The curvature of space is directly related to the matter–energy distribution. In this way the geometry of space, itself of necessity universal, is intimately related to another universal thing, gravitation.

This extremely elegant solution yields all the remarkable effects of the flat-space approach. Unfortunately, all the so-called tests of general relativity so far carried out are really tests only of the principle of equivalence rather than of geometry. Both theories predict essentially the same phenomena. The real difference involves a basic philosophic approach and the argument turns on whether acceleration is something absolute, or like uniform motion, completely relative. If uniform acceleration is absolute, then an observer at infinity is indeed in a privileged position, since his physical situation is quite different to that in a gravitational field. But if acceleration is purely relative, like uniform motion, then the laws of physics, and these include the constancy of the velocity of light, are the same to all observers, whatever their position in a gravitational field, or whatever their acceleration.

Look at the problem the way Newton did in his *Principia*, published in 1687. The 'natural' motion of a body is for it to travel in a straight line with uniform velocity. If we refer this motion to a rotating frame of reference we have to invent two 'inertial' forces, one directed outwards from the axis of rotation, the other directed tangentially. The outwardly directed force is the centrifugal force, the apparent force which assails one when in a car speeding round a corner, and the tangential force is the Coriolis force, the force that makes winds on the Earth follow circular paths more or less parallel to the isobars. Newton considered the following experiment (Figure 38). Half fill a bucket of water, suspend it from a hook, twist the bucket round and round and let go. The bucket rotates, and very soon this rotation, by viscosity, communicates itself to the water. As a result, the surface of the water becomes curved, concave upwards. Clearly the centrifugal force pushes the water to the sides and, since it is virtually incompressible, the water piles up. For Newton this meant that an absolute distinction could be made between rotating and non-rotating systems. In a

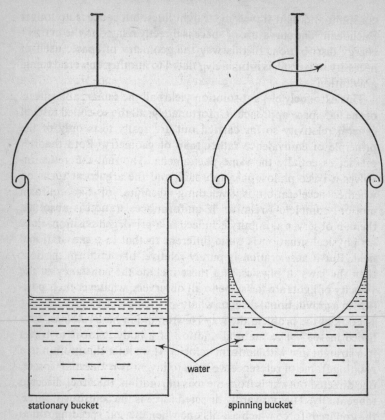

Figure 38. Newton's experiment.

rotating system the surface of water is curved; in a non-rotating system it is flat.

In support, one can point to the fact that the earth bulges at the equator, and so on, all of which points to there being tangible physical effects associated with rotation, effects which would disappear if the earth had never rotated. Surely, then, rotation is not a relative thing, but an absolute thing.

Yet it is difficult to give meaning to absolute motion, as Bishop Berkeley was the first to point out. How could we tell that one sphere rotated around another, if there were no background reference frame of the fixed stars, that is, the rest of the universe? Surely only motion relative to this background has any meaning. This can be accepted happily, without destroying the concept of absolute rotation, if it is assumed that the reference frame of the rest of the universe has absolutely zero rotation. To Ernst Mach, in 1872, this was unacceptable. For Mach only relative motions existed.

But if this were so, then keeping the bucket still and rotating the fixed stars to give the identical relative motion should make the surface of the water curved exactly as before, even though the water was not moving in the accepted sense. Equally well, the equatorial bulge of the earth would appear, if the earth were held still and the rest of the universe rotated. The fact that nothing in our existing laws of nature predicts this behaviour means that our laws are incomplete. In general, all inertial properties of an object are determined by the existence of all other bodies in the universe. This is known as Mach's principle. Einstein accepted it fully in his theory of general relativity, and was led to the concept of curved space–time. But no one has yet produced a fully satisfactory theory of inertial mass based upon the influence of matter in the rest of the universe, although Sciama has offered a simple model.

In Sciama's view inertial forces are due to a new component of the gravitational interaction which is proportional to acceleration. This new force is entirely analogous to the force exerted by an electromagnetic wave which, of course, arises out of the acceleration of a charge. These gravitational-inertial waves are emitted by a body whenever it is accelerated. Suppose we look at the forces experienced by a body falling freely towards the Sun (Figure 39). If acceleration is truly a relative quantity we should be able to describe the situation satisfactorily by taking the body to be at rest and everything else accelerating in the opposite direction. Thus the same physical situation should pertain, if the sun were accelerating towards the body, and if the rest of the universe were subject to this same acceleration. Look at the forces acting on the body. There is still the normal gravitational attraction which will tend to make the body move towards the sun. That has not

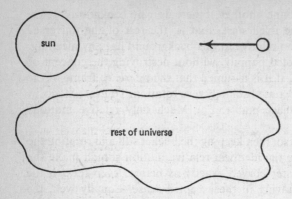

(a) body accelerating in sun's gravitational field.

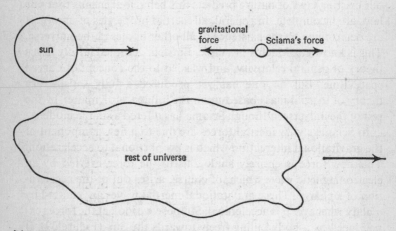

(b) body stationary, everything else accelerating.

Figure 39. Relative acceleration.

changed. But if the body is to remain stationary, this attraction has to be opposed by another force. This is Sciama's new force which is proportional to acceleration and, like the analogous electromagnetic one, inversely proportional to the distance. Both the sun and the rest of the universe exert this new force since they are now accelerating. But the amount of matter at a given distance increases, the further outwards we go. Following the surface area of a sphere, it increases with the square of the distance outwards. Thus the increase in the amount of matter as we go outwards more than compensates for the weakening of Sciama's force with distance. Consequently the major contribution is made by the many, most distant, galaxies. This force just balances the sun's gravitational attraction, if a certain condition on the total mass M and radius R of the universe, is met. The condition is that $2GM/Rc^2 = 1$, where G is the gravitational constant and c is the velocity of light. It is interesting that this condition is one which keeps cropping up in general relativity, and one which defines the gravitational radius of the universe, as we discussed before.

If acceleration is relative, rather than absolute, then the inertial mass of a body is a measure of the force exerted by the rest of the universe when you push the body. What a remarkable idea, that when you accelerate into a run, your muscles are fighting the influence of galaxies scarcely visible even with the most powerful telescopes!

Attempts, like Sciama's, to describe a cosmological origin to inertial forces are to be contrasted with the search, outlined in the first half of this chapter, for the origin of inertial mass, pursued through quantum-field theory. The one attempt must speculate upon the existence of a hitherto unsuspected influence of distant matter in order to save the idea of the relativity of motion, in its general sense; the other must push our concepts of microscopic matter down to infinitesimally tiny dimensions in order to calculate the self-energy, and hence inertial mass, of elementary particles. In these fundamental problems matter in the largest and smallest scales is involved, and one cannot avoid the feeling that there exists a strong skein in the substratum of nature, which connects the size of the universe to the size of elementary particles. One day this skein will be unravelled. But the theory which does so will have to answer another question as well as the problem of inertia. That question is one we have not mentioned, since nothing positive as yet exists to point to its answer. It is, why do the elementary particles have the masses that are observed?

CHAPTER 9
CHANCE

Chaos umpire sits,
And by decision more embroils the fray
By which he reigns: next him high arbiter
Chance governs all.

Milton: *Paradise Lost*

IT is all very well having a beautiful system of ideas to apply to the case of one thing interacting with another thing, but the real world is far more complex than that. Think of the myriad interactions taking place at this moment between the atoms composing the paper of this book. Think that in a cubic centimetre of any solid, there are roughly 10^{22} atoms (a quite unimaginable number), each one interacting electromagnetically with every other to varying degrees, making a total of 10^{44} interactions in all. Think that even with more rarefied matter, such as the air we breathe, there are around 10^{22} molecules in every litre which collide with one another and with the boundaries which contain them. Large numbers of things in continual motion, which itself is continually changing; such appears to be the essence of real matter about us. How can such a chaotic state be described?

There are, of course, simple properties which a chunk of matter possesses. It has a certain volume in space, it has gravitational and inertial mass, it has (in principle) a countable number of particles and a countable number of vibrational modes. It may have an electric charge or it may be magnetized. It may, in addition, possess motion as a whole, as a tennis ball projected over the net has, or a spinning top has. Our ideas of identity, space and time, energy and momentum, and the various types of interaction, are quite good at dealing with properties like this. But what can they say about the chaotic internal motion brought about by everything interacting with every other thing inside our chunk of matter?

It is essential that something be said about it, if only because it can be observed. Small particles, suspended in a liquid, move about at random, executing the so-called Brownian motion (named after the discoverer). Such motion reflects the continual bombardment by the molecules composing the liquid. Gas exerts a pressure on the walls of its container. Any tennis player knows the force which a moving tennis ball exerts when it bounces off his racquet. Pressure is just a manifestation of bouncing molecules. Listen to the meaningless hiss of a badly tuned radio or television: a hiss or 'noise' whose origin is the random motion of electrons – meaningless, incoherent activity. Good browsing here for the philosopher and cynic; but essential browsing for the natural philosopher, who looks for meaning in chaos – the extraction of information from a background of random events – and for the experimentalist concerned with the limits of accuracy. The all-pervading, levelling play of hazard in the universe, opposing implacably the acquisition of detailed knowledge, must inevitably be encountered and challenged in the quest for meaning.

Perhaps an even more compelling reason for understanding the chaotic state is that the ideas of physics which we have considered so far, and which work beautifully when only a few particles are involved, do not explain several familiar and everyday events. What have they to say about getting hot, or cooling down? When we look at a thermometer and read off the temperature as so many degrees centigrade; what meaning does it have, what exactly is being measured? Why, above all, has the practice of magic to be an illusion and not a reality? For example, why do broken glasses never mend themselves, why does the air in the room never suddenly decide to blow itself into a corner? The point is that the laws of physics, whether they be to do with mechanics, electromagnetism, gravitation, or strong nuclear forces, do not forbid such things. The projector can be run backwards – glasses mend themselves, the air rushes into a corner, explosions implode – and none of the laws need be violated. Energy can remain conserved, and so can momentum. But such things do not occur. Why not? The laws are silent. They are incomplete.

What makes them incomplete is the lack of ideas associated with large numbers. Physics is always particularly easy when properties are extreme – absolutely hard, perfectly elastic, extremely weak or overwhelmingly strong. In such cases ideas are very simple, theories produce neat expressions, and so on. It is only in the murky, grey areas, where properties lie midway between poles, that things get messy. The physics we have been dealing with hitherto is nice and simple when there are only a few particles. Can we hope that when the number of things gets large we reach an opposite pole where simple patterns emerge? We can. If order and certainty is one end of the spectrum, may we look for simple ideas at the opposite, where absolute disorder and ignorance reign? We may. There exist laws governing the behaviour of large numbers – the laws of probability.

The concept of probability has to rank with the other basic ideas of physics as an equal alongside those of space, time, energy and momentum. But, although it is easy to grasp what space is and what time is, and not difficult to learn to know what energy and momentum are, and how they differ from one another, the ideas about chance sometimes seem difficult to appreciate and appear occasionally to offend intuition.

It is acceptable that in a toss of a coin a head is just as likely as a tail. It is acceptable to define the mathematical probability of a head as a half and of a tail as a half. There are a total of two possibilities and the sum is unity, representing absolute certainty that either a head or a tail will turn up. If the coin were weighted in some way to favour heads, there would still be only two possibilities, but the numerator of the mathematical probability of a head would have to be a number greater than one (but less than two) to describe the asymmetry. The possibilities would not have equal weight. But let us suppose they do have equal weight. What is the probability of throwing two heads in two tosses? Two heads is one possibility, two tails is another. Yet another is a head on the first toss and a tail on the second, and yet another is the opposite of this, namely a tail on the first toss and a head on the second. In all, then, four possibilities, and two heads is only one of these (Figure 40). The probability of two heads is a quarter, which we could have arrived at by multiplying the probability of a head in the first

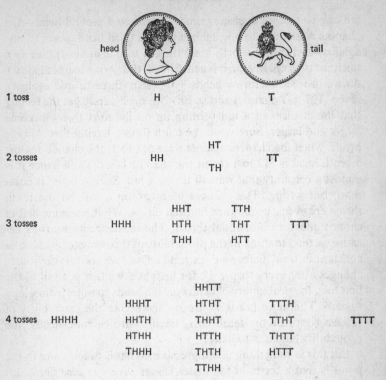

Figure 40. Possible results of coin-tossing.

toss by the probability of a head in the second. So the probability of ten heads in ten tosses is half multiplied by itself nine times viz $(\frac{1}{2})^{10} = \frac{1}{1024}$. If we cared to work it out we would indeed find that there are 1024 possibilities, only one of which is the ten-head situation.

If we set out to throw ten heads in ten tosses the chances against our succeeding are 1023 to 1. If a bookmaker offered these odds, neither he nor the punter would gain in the long run. (Actually, we would be lucky to get 100 to 1 from any self-respecting bookie, what with overheads, wages, taxes, etc.) Imagine starting our ten tosses. We throw a head. Well, that is not amazing since the odds are one to one, or evens, against. We throw a second head – the chances were three to one against this – a third head (seven to one against) and then a fourth head (fifteen to one against). Surely a tail must turn up. But no, it is a fifth head (thirty-one to one against). As we proceed to throw heads – six (sixty-three to one against), seven (127 to 1 against) and so on – we must surely get the feeling that the chances of a tail turning up on the next throw become larger and larger. Surely, on the tenth throw, having thrown nine heads, when the chances against were 511 to 1, the chance against a tenth head is 512 to 1 (to fit the 1023 to 1 odds) and hence it is almost a certainty that we will throw a tail. Such a belief is common, but wrong. The chances of throwing a tail on the tenth throw are evens, as they are on each throw. What has gone before cannot affect an individual throw. The misconception arises because we tend to think of the probability of throwing ten heads as a constant in time. Before we start, it is. Once we start, it obviously changes after every throw. If, for instance, we throw a tail in the first toss, the probability of achieving ten heads changes from $\frac{1}{1024}$ to zero. The more heads we throw, the higher the probability of success, until at the tenth toss, having thrown nine heads, the probability has risen to half.

But this is something of a digression. What is much more to the point is that a series of ten tosses is *most likely to give five heads and five tails*. This is because there are more ways of arranging five heads and five tails in various orders through the ten tosses than there are for any other combination. The series of ten heads is a freak. The most probable result is the one consisting of equal numbers of heads and tails. Here is an order coming out of haphazard events, in the shape of the most probable result. It is the most likely state of affairs which concerns us.

Suppose we attempt to simplify coin tossing (conceptually, that is) by regarding it as a mechanism for producing, on average, equal numbers of heads and tails, and ignoring all other possibilities. How wrong would that picture be? In a two-toss run the most probable number of heads to be expected is one. But the chances of no heads or two heads are quite high. Roughly, the expected number may be written 1 ± 1 (one, plus or minus one), corresponding to a 100 per cent error. To regard coin tossing as a means of producing equal numbers of heads and tails in this case is extremely inaccurate. It is somewhat better in the ten-toss case. Here the expectation is five heads, but we would not be surprised to get six or seven or four or three, though we would think it unlucky to get nine or ten or one or two. The expected number may be written roughly 5 ± 2 (five, plus or minus two), i.e. a 40 per cent error. The error is less, but still appreciable. If we work out mathematically what error to expect for large numbers, we find that it goes as $\dfrac{1}{\sqrt{N}}$; where N is the number of tosses. In the hundred-toss case the error is down to 10 per cent. If we jump incredibly to some 10^{22} tosses, the error is 10^{-9} per cent, or one thousand millionth of a per cent. No one is going to cavil, if we ignore everything but the most probable situation, with numbers like this.

The sheer size of numbers like 10^{22} produces its own simplicity. But what is the nature of the most probable result on which we have now to focus attention? By definition it is the one that can be achieved in more ways than any other. It contains the largest number of possibilities. In the sense that the situation consisting of 10^{22} heads represents maximum order and certainty of knowledge, the most probable state represents maximum disorder and ignorance. In another sense the all-head situation allows no freedom of choice to the individual, whereas the most probable state contains the greatest degree of freedom. In the all-head situation a head that wishes to become a tail (if such a desire can be imagined) threatens the state itself; but a similarly inclined head in the most probable state has many possibilities of persuading one tail to become a head, and so satisfy his ambition without destroying the state. The most probable state is therefore the most stable, because it is least ordered. All manner of sequences can yield equal numbers of heads and tails. All manner of complexions are compatible within the framework defining the state. It is tempting to draw analogies with human society at this point, but we will be strong and refrain.

Two factors, then, characterize the most probable state. One is the distribution of components, e.g. equal numbers of heads and tails, the other is the degree of disorder. What can be taken to be the measure of *disorder*? The simplest is just the number of ways of getting the most probable distribution of heads and tails. Let us call this number C. In the two-toss case $C = 2$, a head then a tail or a tail then a head. In the ten-toss case it turns out that $C = 252$, that is, there are no less than 252 ways of getting five heads and five tails. In the 10^{22} toss case, C is astronomical (roughly, two raised to the power 10^{22}). That C rises approximately exponentially with toss-number would in itself not be too bothersome, but it does offend our psychological feeling somewhat in the same way as the lily-clock did in chapter 4. Going from a two-toss to a ten-toss case we feel that the disorder should be increased by a factor nearer to five rather than 126. Furthermore, consider a composite situation consisting of two ten-toss cases. For every one arrangement in one of the ten-toss cases there are 252 arrangements in the other, so the total number in the composite system is 252×252. This again offends against what we naturally expect. The disorder is increased by a factor of 252, whereas we feel it ought to be merely doubled. And there is yet another criticism. In a situation where $C = 1$ there is no disorder. Surely the number denoting this condition should be zero.

Fortunately there is a mathematical function of C which overcomes all of these objections. It is the natural logarithm of C, written lnC, and we can look this number up in tables of natural (Naperian) logarithms (Figure 41). Taking logarithms reduces exponential variations to linear ones, allows us to add disorders rather than multiplying them, and ensures that when $C = 1$ the result is zero. This appears altogether a more satisfactory measure of disorder, and indeed it works out to be so in practice. The disorder associated with the most probable state in each of the two-toss, ten-toss and 10^{22} toss cases is then respectively $ln2 = 0.69$, $ln252 = 5.53$, $ln = 10^{22} = 6.9 \times 10^{21}$.

And now, to make a bridge from tossing coins to the real world, it is vital to make a point concerning the composite state consisting of two ten-toss runs. The point is extremely fundamental and essential to the understanding of why magic is hard to come by. The

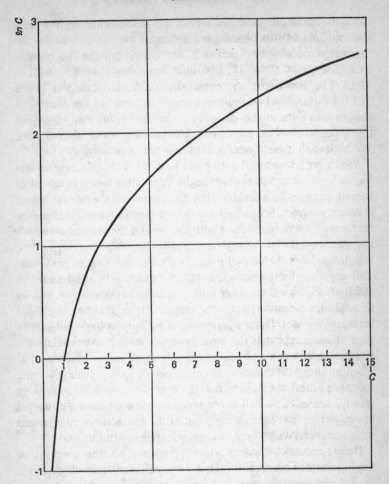

Figure 41. Natural logarithmic curve.

disorder in one of the runs is, as we have just noted, $1n252 = 5\cdot53$. In the composite state it is $1n(252 \times 252) = 2 \times 5\cdot53 = 11\cdot06$, just twice what it is in the single case. Now it is interesting to ask what it becomes, if we mix the two runs together and have, in effect, a twenty-toss case. How does the twenty-toss case compare with the two ten-toss composition? Clearly, the number of heads does not change – ten in one, and two lots of five in the other. If, however, the two lots of five become ten arranged over twenty tosses the number of possibilities increases. Working it out for the twenty-toss case gives $C = 185,000$ and hence the disorder, $1nC$, is 12·13. The disorder in the composite, unmixed, state was 11·06. Thus, by allowing two ten-toss systems to 'mix' and become one twenty-toss system the disorder is increased, without changing the appearance of the most probable state, i.e. ten heads, ten tails. We believe that real systems behave in the same way.

Whatever interactions with other bodies a real body may undergo, the result is an increase in the disorder of the total system, or at best no change. This is essentially a statement of the second law of thermodynamics. Broken glasses do not mend themselves, because that would mean interacting with the air and the ground on which they lie, in such a way as to decrease disorder. This is not forbidden, it is just so improbable that it can be discounted. One might just as well expect a handkerchief, flapping aimlessly in the breeze, to suddenly pull itself together and, vibrating harmoniously, render us a choice selection from the classics. It could, but it almost certainly will not. Quite the reverse; it is likely to flap aimlessly to rags. If we accept that the most probable state is overwhelmingly to be expected in a physical system, then we have to accept that when mixing or interaction between two or more systems occurs, there is a practical certainty that the total disorder will increase or stay the same, but it will never decrease. Here we have a powerful law of nature that arises simply out of the character of chance and large numbers, which explains many features of our universe.

If the concept of disorder remains the same, whether we think of tossing coins or concern ourselves with real, many particle,systems, the description of the most probable state in the real world is a considerably more complex affair. A real system consists of huge numbers of particles of various sorts, in individual and collective motion, occupying a certain amount of space for a certain length of time. The description of the most probable state is therefore obviously a much more formidable task. Nevertheless, the ideas which we can bring to bear on the problem are simple and straightforward.

First of all we invent a simple concept which gets rid of the time element. Imagine our real system to have remained undisturbed, in physical contact with its surroundings, for a long period, so that its properties have become independent of time – such as putting a chunk of meat into the freezer and leaving it there to reach and remain at the temperature inside. We say that the system is then at thermodynamic equilibrium with its surroundings, and if we confine our attention to systems at equilibrium we simplify our task considerably.

Next we specify the number and type of particles in the system and the volume which the system occupies. That takes care of identity and space. Then we have to enumerate and define all the dynamical states which are possible. This means specifying every value of momentum a particle or a wave can have (which is actually easier than it sounds). Since we suppose we understand the physics of the system, the relation between momentum and energy is known in all cases, and so the possible energy states of the system are also thereby defined. We can imagine these states as so many boxes into which we put particles or quanta, and so we can contemplate all the possible ways in which this may be done and eventually pick out the most probable distribution, the one with most disorder.

But to get a precise answer we need to specify something more. We need to define two quantities. One is the total momentum. In a stationary system there will be as much momentum of particles and quanta in one direction as in the opposite. Since momentum is a directional quantity the total will, under these circumstances, be zero. A consideration of moving systems would take us into the esoteric field of relativistic thermodynamics, so let us stick to nice simple stay-at-home bodies. The other quantity which has to be defined is the total energy. And this has a definite magnitude, being made up of the sum of all kinetic energies and vibrational energies. It is, in fact, the thermal energy or, more plainly, the amount of heat in the system. Heat is nothing more or less than energy of randomized motion. Given the total heat content, we can work out the most probable distribution, and hence derive certain average quantities, like the average kinetic energy of a particle or the aver-

143

age amplitude of a wave; and we can also work out the disorder. Increasing the heat content increases the number of ways of distributing particles and quanta among the allowed energy states, and that means greater disorder. Conversely, reducing the heat content reduces the disorder. Thus the total heat content appears to be a vital quantity, which has to be specified for every system.

Unfortunately, the heat content of a body turns out to be a very poor concept in practice. When we walk into a cold room and comment on the fact that it feels cold, we are not really making a statement about the heat content of the room. A large, cold room could have a higher heat content than a small, hot one. We are making a remark about the temperature. In the study of heat we do not observe heat to flow always from bodies of higher heat content to those of lower content. Which is just as well, since our cold room may pour heat into the electric fire rather than the other way around. The quantity which determines the direction of heat flow is the difference between temperatures, and not differences between total energies. Heat always flows from bodies at high temperature to bodies at low temperature. A body is hot if its temperature is high. Temperature is the vital quantity, not heat content.

But what is temperature, and how is it different from heat content? We have a direct appreciation of temperature through our senses. We make measuring devices, known as thermometers, in order to quantify it. Of all properties which can be associated with the thermal behaviour of bodies, temperature is the most meaningful. And yet, in our statistical model there has not been a hint of a concept that could explain what temperature is. We have talked about systems in terms of numbers and types of particles (atoms, molecules, electrons), types of quanta (phonons, photons), energy states, total energy content, the most probable distribution, and disorder. All aspects have been covered. Since temperature is not one of these explicit quantities, it must be associated with some relationship between them. The idea of temperature must be already implicit in our model. We just have to dig it out.

First of all, temperature must be intimately related to heat content. Although the comparison of heat contents of different bodies is a poor indication of how much hotter one body is relative to another, it is nevertheless true that in the case of a single body of fixed volume the addition of energy raises its temperature. So it is true that the greater the heat content the higher the temperature for the case of any one system. But we can arrange to double the size of the system without changing the temperature. The heat content will have doubled in the process, but this is not reflected in the temperature. Temperature is not dependent upon size, but total energy content is. This suggests that temperature is something to do with total energy content divided by some quantity which is dependent upon size. In other words, it is associated with some sort of average energy characteristic of how hot the system is. But what sort of average energy?

Now the total energy is just the sum of the kinetic energies of all the particles plus the energy in all the vibrational modes. It is tempting to form a characteristic energy by dividing the total energy by the number of particles and vibrational modes. But this does not work. One reason is to be found in the mechanical difference between purely translational motion and oscillatory motion. Technically speaking, translational motion has three degrees of freedom (along, sideways, and up) whereas oscillatory motion has two (backwards and forwards). If we take this into account then indeed the characteristic energy we get behaves like temperature, provided that quantum effects are negligible. In the régime where classical physics works, it is utterly valid to regard temperature as a measure of the average kinetic energy of a particle or of the average energy in a vibrational mode. Now energy is measured in joules, and temperature is measured in degrees absolute, being defined in terms of the pressure of an ideal gas. When the pressure of an ideal gas is zero the gas is at zero temperature. On the absolute temperature scale ice melts at 273° and water boils at 373° (roughly), so a degree on this scale is identical to a degree on the centigrade scale. The conversion factor which relates energy to temperature is known as Boltzmann's constant, symbol k, equal to $1 \cdot 3805 \times 10^{-23}$ joules per degree. The characteristic energy is then denoted kT, where T is the absolute temperature. The average kinetic energy of a particle turns out to

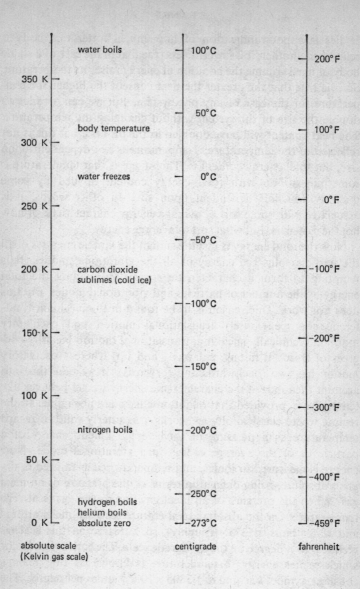

Figure 42. Temperature scales.

be $\frac{3}{2}kT$ and the average vibrational energy in a mode is kT. The ratio just reflects the ratio of the degrees of freedom.

Unfortunately, this scheme does not work in the quantum régime. The simplest way of seeing this is to consider a population of electrons. Electrons obey the Pauli exclusion principle – no two electrons can occupy the same state. As the temperature is lowered, the lowest energy states get filled up and many electrons are thereby forced to remain in higher energy states (Figure 43). At the

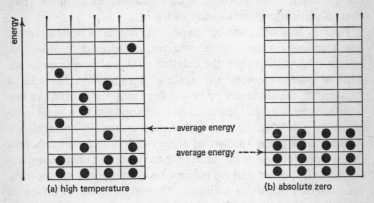

(a) high temperature   (b) absolute zero

Figure 43. Electrons occupying dynamical states. The Pauli exclusion principle allows only one electron per state.

zero of temperature the total energy of the electron population is as low as possible, but it cannot be zero. The average kinetic energy may in fact be quite high. Clearly, the simple relationship between average energy and temperature completely fails here.

Temperature, then, is not definable in terms of average kinetic or vibrational energies. We need something more general. We tried using the number of particles and the number of vibrational modes to define the size of the system, but this has failed. What remains in our statistical model which would serve instead? The answer is the disorder.

Imagine the following experiment. A gas is contained in a cylinder closed at one end by a rigid wall and at the other by a movable piston. Let a small quantity of heat cause the gas to ex-

147

pand and push the piston along the cylinder a little way. Suppose that all the energy supplied went entirely towards moving the piston, then no rise of temperature would be involved. Because the gas now occupies more space, there are more possible ways of achieving the most probable state. In other words the disorder has increased by a definite amount. A measurable quantity of energy added has produced a measurable increase of disorder. There exists therefore a characteristic energy which is the amount of energy required to produce a unit increase in disorder. This characteristic energy we identify with kT.

Happily this definition of temperature works in both quantum and classical régimes. In relating temperature to disorder, we isolate its definition from the detailed statistical properties of the various components of the system. Temperature becomes a measure of the difficulty of producing further disorder. At low temperatures a given increment of energy produces a greater increase of disorder than at high temperatures. Conversely, the removal of a given amount of energy at low temperatures produces a greater degree of ordering than it would at high temperatures. Perfect order corresponds to the absolute zero of temperature. Since disorder is intimately bound up with numbers of particles and numbers of modes, it is not too surprising that our definition based upon average energy came close to the truth. Nevertheless, the relationship between temperature and the energy to produce disorder is the fundamental one.

If we multiply the disorder by Boltzmann's constant we obtain a physical quantity with units joules per degree. This quantity is known as the entropy. Temperature then is the energy to produce unit increase of entropy. It says a lot for the science of thermodynamics that the concept of entropy was found to be necessary, long before any detailed statistical models came along. In any thermodynamic process entropy is the quantity that never decreases. Such is the content of the second law of thermodynamics. Relating entropy to the concept of disorder is the triumph of statistical physics and renders the second law entirely intelligible.

The role of chance is therefore as vital in physics as the role of gravitation and electromagnetism. It regulates the distribution of energy within large populations of things and lays down fundamental laws which govern the flow of energy from one large population to another. The laws of chance determine that, on average, heat flows from hot bodies to cold bodies, that perpetual motion machines can never work because machines require ordered energy and energy always tends to become disordered, that an arrow of time is defined in terms of increasing entropy. Entropy may decrease locally, for example, in a refrigerator, but in the system at large, for example the refrigerator plus the room in which it operates, the total entropy always tends to increase. We can run machines successfully only because order exists and can be tapped to do useful things. Fortunately, the world is very far from being at thermodynamic equilibrium, and there is a long way to go before disorder approaches anywhere near its ultimate limit. It is unlikely that the situation near maximum disorder will, in any case, allow the existence of highly-organized life forms like ourselves, so the failure of our machines, with little or no order to feed off, becomes somewhat academic.

Actually, the universe is a highly ordered place, even after 10^{10} years of 'running-down'. Matter is very largely concentrated, in the cosmic scale, into galaxies, and not in the least spread uniformly; and there are vast amounts of tappable energy within the hydrogen nuclei, swirling in great star-forming clouds of gas. Nearer home, hydrogen fusion in the sun provides an ordered stream of energy which warms the earth, fuels the vast atmospheric-oceanic, weather-producing, heat-engine, that arranges energy storage in the form of oil, gas, coal and other fossil fuels, and encourages the highly improbable chemical organizations of living things. Disorder may win in the end, but the rear-guard action of certain ordered systems can be very fierce indeed. That is because they have defences.

One defence which ordered systems use to protect themselves from disorder is the energy barrier. Fossil fuels will not burn until heated. Some energy has to be applied to initiate the heat-producing chemical reaction of burning. Explosives need a detonator; again, a reluctance to react has to be overcome by an input of energy. Hydrogen nuclei will not fuse until energy is provided in sufficient quantity to overcome the electrostatic repulsion between the two protons. Attempts to overcome this proton barrier in a controlled way eat up sizable chunks of national incomes all over the world. (It is just as well that this particular barrier is so formidable, otherwise the arrow shot by chance would have travelled a good deal faster.) Moreover, ordered systems, by their very nature, dig energy wells to contain their various parts, and to protect their organizations against the universal levelling process. The deeper the well, the more stable is the system. Every ordered system has an energy well and some have, in addition, the protection of a barrier (Figure 44). To destroy such a system not only does one have to haul it out of a well, one also has to drag it over a mountain. Energy wells and barriers defend against disorder.

And very often they can turn disorder to advantage, and get it to create order, locally. For, suppose we have a measure of disorder in a system – say, a whole lot of particles jiggling about. Suppose there is the possibility of two or more particles combining and digging themselves an energy well, so forming a little ordered system – but they have to overcome an energy barrier to do so. If the jiggling is weak, that is, the temperature is low, very few particles will be able to climb the barrier. But now raise the temperature; increase the jiggling and, incidentally, the disorder. More little ordered systems will be formed! Ironically, overall dis-

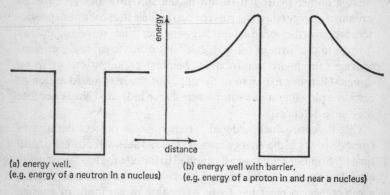

(a) energy well.
(e.g. energy of a neutron in a nucleus)

(b) energy well with barrier.
(e.g. energy of a proton in and near a nucleus)

Figure 44. Energy wells and barriers.

order produces, here and there, local order! Of course, if we go on increasing the temperature the little ordered systems will eventually jiggle themselves to bits, but nevertheless there exists a range of disorder which not only permits, but actually encourages, the formation of local order. Life itself must owe its origin to such a paradox.

CHAPTER 10

MYSTERY

Woe unto them that join house to house,
that lay field to field, till there
be no place.

Bible: *Isaiah*

We shall not cease from exploration
And the end of all our exploring
Will be to arrive where we started
And know the place for the first time.

T. S. Eliot: *Little Gidding*

LOOKING back over the exploration of things, and space, and time we cannot help, I think, being struck by the thought that what is so remarkable is not how much we understand nature, but how much we don't. And it is not just that, inevitably, there is a fell-walking effect – once you have got to the top of one rise you see another, previously hidden, rise before you. Of course, that effect is bound to be an indigenous feature of any exploration, and makes getting to the top of a rise a rather exciting affair. There can be few things more exhilarating than seeing a way to the top, getting there, and viewing, for the first time, a panorama hitherto unsuspected. But the mystery of what lies over the brow is not the only one, though it may be the most compelling. There is also the intrinsic mystery of the territory already explored, appearing as caves and pot-holes in the landscape, reminding us of deep, inaccessible, places, about which we know nothing.

Look, for example, at the chunk of rock we contemplated as a missile at the beginning. It is just one of countless other chunks of matter we find about us. We know it is built up of certain molecules out of the countless possible molecules that exist. We know the molecule is made up of atoms, and there are only about a hundred different sorts of atom. We know the atom is made of electrons, protons and neutrons, and we think these particles may be made of truly elementary particles called quarks or partons (or what-you-will-ons). Brow upon brow has been surmounted. The atomic fell-walking has led us to the formidable peak of the quark. Is this the summit? Or is there a whole mountain range yet to come?

We observe the mutual attraction of chunks of matter, call it gravitation, and describe what happens in terms of gravitational charge, space, and time. We observe the phenomena of electricity and of magnetism and describe what happens in terms of inertial charge, electric charge, permittivity, and of course space and time. We realize that space and time are linked indissolubly to the operations by which distances and times are measured, and we choose a space-time framework defined by electromagnetism, through the velocity of light. We discover the weak interaction – a force which makes neutrons decay; and the strong interaction – a force which holds protons and neutrons together in the nucleus. More strangely we find elementary particles behaving as if they possessed a limited freedom to ignore the classical laws of cause and effect, and we describe such behaviour in terms of probability waves and the quantum of action. We see the laws of chance, familiar to macroscopic matter, applicable to the behaviour of a single particle, and find that familiar, stand-by, conservation laws, those of energy and momentum, hold only in a statistical way. To encompass this behaviour, and to allow for particle-anti-particle creation and annihilation we invent the quantum field. With it we can derive the energy levels of an electron in a hydrogen atom with impressive accuracy. It works, and to that extent nature is understood. A lot of ground has been covered, and brought under dominion. But as we walk around, admiring the view, let us beware of the caves and pot-holes.

One has the strong suspicion that underlying the well-mapped fields of nature is a whole labyrinth of interconnections, only very dimly appreciated. Why is inertial mass apparently identical to gravitational mass? Is there a passage connecting the inertial behaviour of submicroscopic particles with the totality of matter in the universe? Why have the elementary particles the masses that are observed? Here is a prominent cave, visible from every region of the physical world, the Cave of Mass. It beckons the intrepid caver and promises to lead him to subterranean connections between black holes and particles, the very big and the very small.

Near by is the Pot-hole of Infinite Zero-point Energies, a daunting challenge. Usually people go round it one way or another, but it constitutes a navigational hazard and it really ought to be filled in. Quantum fields have this disturbing feature of zero-point fluctuations, continually creating and annihilating their own particles. The energy in any one dynamic mode is never zero, never falls below the zero-point energy of the mode. The number of possible dynamic states a particle may have appears to be infinite, and therefore the total amount of energy, even in a vacuum, appears to be infinite. Energy produces gravitational fields, in this case infinitely large ones. Something is very wrong. Our concept of dynamic modes is incomplete at high energies. There has to be something that cuts off the sum and prevents infinities. Perhaps a cut-off may come about because space is essentially atomic in nature, rather than continuous. There may exist an elementary length which, since time and space are intimately related, would imply an elementary period. Like a crystal, space would have a lattice structure with a unit cell of sub-nuclear dimensions; like a clock, time would tick. Pure speculation, naturally, but worth looking at seriously.

While we are on the topic of space and time we must point out a cave that few would attempt to explore seriously without considerable qualms. The daunting object is the Dimensions Cave, opening into Who Knows What with its lure of 4-dimensional space (why stop at 4?) and multi-dimensional time. Marginally more popular may be the Time Cave, situated near by, which offers the options of gravitational time, neutrino time, and nuclear time, to those intrepid enough to question the applicability of the much used electromagnetic time. And to complete this travel guide to the unknown we should mention the cave of Unification, in which gravitation and electromagnetism hopefully are to be found integrated within a unified field theory, which must eventually encompass the weak and strong interactions, and may explain why certain dimensionless quantities formed from the fundamental constants are what they are.

The latter are the solid, quantitative, mysteries of nature and they stand like pillars in the landscape. Naturally, the magnitudes of the fundamental constants themselves are not the important things, since they depend upon what standards of length, time, charge, mass, etc., are chosen. In metres per second the velocity of light, for example, is (roughly) 3×10^8. If we choose, instead, centimetres per second, its magnitude increases to 3×10^{10}. Only dimensionless quantities carry significance, since they are completely independent of choice of units and are therefore pure numbers. The most prominent dimensionless entity, one which effectively measures the strength of the electromagnetic interaction, is the fine-structure constant $e^2/2\varepsilon_0 hc$, where e is the elementary electric charge, h is Planck's constant, ε_0 is the permittivity, and c is the velocity of light. Its value, a pure number, is very nearly $\frac{1}{137}$. If we define the nuclear binding in terms of a 'charge' q, then the analogous quantity for the strong interaction, $q^2/2\varepsilon_0 hc$, turns out to be about unity. Doing the same thing for the weak interaction we get a number of the order 10^{-13}, and for gravitation a tinier number of order 10^{-39}. Thus the relative strengths of the four fundamental interactions go roughly as 1, $10^{-2}, 10^{-13}, 10^{-39}$.

These numbers, and others that we have mentioned in earlier chapters, such as the ratio of greatest and smallest lengths (10^{-40}), the ratio of longest and shortest times (10^{-40}), the number of protons and neutrons in the universe (10^{-78}), have ultimately to be explained. At present they remain mysterious formations in the physical landscape.

No doubt further caverns are waiting to be discovered by the fell-walker in his hunt for quarks, or, at the other end of the distance scale, in his quest for black holes, and in his investigation of quasars, whose prodigal rate of energy production is scarcely understandable in terms of physics as we know it. But caving and fell-walking are not the only activities. There is the vast task of prospecting fields whose general geography is understood, fields whose mysteries stem rather from the complexity of macroscopic matter than from fundamental structure. Think of the rich veins of ore to be mined from the solid-state field; the curious seams of the low-temperature world; the startling versatile outcrops of the semiconductor and insulator fields; the polymer intrusions. Think of the liquid crystals, the volcanic plasmas; the slanting, deceptively ethereal, laser beams; some territories with jewels lying on the surface waiting to be picked up, some with jewels deeply hidden, none without jewels; the rich fields of chemistry and the vast sub-continent of biology. In them all the basic structure is established, the caves and pot-holes are scarcely visible in the far-off foot-hills and mountains, and there is overwhelmingly the busy, bewildering complexity of inanimate and animate matter to explore.

The mystery of things, the mystery of relations and inter-relations and the mystery of complexity, claim our wonder and our involvement, whether we be fell-walkers, cavers, or prospectors.

APPENDIX 1

FUNDAMENTAL PHYSICAL CONSTANTS
(1974)

Quantity	Symbol	Value	Units
Speed of light in vacuum	c	$2 \cdot 997924580 \times 10^{8}$	ms^{-1} (metres per second)
Elementary charge	e	$1 \cdot 6021892 \times 10^{-19}$	C (coulomb)
Permittivity of vacuum	ε_0	$8 \cdot 854187819 \times 10^{-12}$	Fm^{-1} (farads per metre)
Planck's constant	h	$6 \cdot 626176 \times 10^{-34}$	Js (joule-second)
Gravitational constant	G	$6 \cdot 6720 \times 10^{-11}$	Nm^2kg^{-2} (Newton-square metre per square kilogram)
Boltzmann's constant	k	$1 \cdot 380662 \times 10^{-23}$	JK^{-1} (joules per degree absolute)
Bohr radius	a_0	$5 \cdot 2917706 \times 10^{-11}$	m (metre)
Rest-mass of electron	m_e	$9 \cdot 109534 \times 10^{-31}$	kg (kilogram)
of proton	m_p	$1 \cdot 6726486 \times 10^{27}$	kg
of neutron	m_n	$1 \cdot 6749544 \times 10^{-27}$	kg
Compton wavelength of electron	λ_{Ce}	$2 \cdot 4263089 \times 10^{-12}$	m (metre)
of proton	λ_{Cp}	$1 \cdot 3214099 \times 10^{-15}$	m
of neutron	λ_{Cn}	$1 \cdot 3195909 \times 10^{-15}$	m
Magnetic moment of electron	μ_e	$9 \cdot 284832 \times 10^{-24}$	JT^{-1} (joules per tessla)
of proton	μ_p	$1 \cdot 4106172 \times 10^{-26}$	JT^{-1}
of neutron	μ_n	$9 \cdot 6629738 \times 10^{-27}$	JT^{-1}

APPENDIX 2

ABBREVIATIONS

Decimal fraction or multiple	Prefix	Symbol
10^{-18}	atto	a
10^{-15}	femto	f
10^{-12}	pico	p
10^{-9}	nano	n
10^{-6}	micro	μ
10^{-3}	milli	m
10^{0}	—	—
10^{3}	kilo	k
10^{6}	mega	M
10^{9}	giga	G
10^{12}	tera	T

INDEX

Potential,
  electric, 116
  gravitational, 127, 128
Power, 94, 109
Pressure, 37, 94, 135, 145
Probability, 32, 102, 103, 136, 138
  mathematical, 136
  waves, 32
Properties,
  chemical, 17
  elastic, 13, 18
  electromagnetic, 23, 27, 83
  holistic, 19
  mechanical, 16, 17, 23, 27
  solid state, 57, 112
  thermal, 16, 17
Proton, 14, 27–30, 33–8, 53, 83–6,
    95, 103, 112–15, 127, 128, 150,
    153, 155
  gravitation radius, 127
Psychology, 15
Pulsar, 37

Quantum, 14, 34, 38, 111, 113, 143–8
  of action, 32, 33, 38, 53, 58, 59,
    100, 101, 104, 105, 107, 111, 115,
    121, 153
  energy, 109
  field, 108, 109, 111, 133
  mechanics (theory), 100, 102, 103
  oscillation, 33
  particle, 33, 34, 58
Quark, 35, 118, 153, 155
Quasar, 155

Radar ,24, 57
Radiation, 24, 94
  cosmic, 37
  electromagnetic, 32, 33, 57, 62, 97,
    104, 109
  γ (gamma), 24, 33
  infrared, 24
  pressure, 94
  sun's, 94
  ultraviolet, 24

  visible, 24
  X-, 24
Radioactive atom, 59
Radio astronomers, 37
  waves, 24, 33, 37, 58, 76
Radius, 37, 47, 48
Random walk, 55
Reflection, 19, 60, 62, 78, 94
Refraction, 19
Regular solid, 12, 13
Relativity, 25, 67, 71–5, 126, 128, 131,
    133
  contraction, 69, 72, 126, 127
  general theory, 128, 129, 131, 133
  mass increase, 126
  rotation, 69
  special theory, 69, 96, 100, 116,
    120, 121, 126
  time-dilation, 69, 72, 74, 126, 127
Resistance, electrical, 112
Response time, 57, 58
Rotation, 90, 102, 130, 131

Schrödinger, 103
Sciama, 131–3
Sciences, 15, 45
Second, 54–7
  mean solar, 57
Seismic waves, 19
Semi-conductor, 23, 27, 156
Shadow zone, 19
Shear waves, 19
Short-range order, 16
Silver nitrate, 85
Simple things, 11, 13, 17, 18, 29, 37, 38
Sinusoidal waves, 20, 76
Size, 40, 41, 52, 53
Slit, 21, 25, 30
Sociology, 15
Solar system, 13, 16, 59, 127
Solid, 14, 17–19, 23, 28, 32, 57, 58,
    134
  amorphous, 16
  crystalline, 16
  state, 57, 58, 156

MORE ABOUT PENGUINS
AND PELICANS

Penguinews, which appears every month, contains details of all the new books issued by Penguins as they are published. From time to time it is supplemented by *Penguins in Print*, which is our complete list of almost 5,000 titles.

A specimen copy of *Penguinews* will be sent to you free on request. Please write to Dept EP, Penguin Books Ltd, Harmondsworth, Middlesex, for your copy.

In the U.S.A.: For a complete list of books available from Penguins in the United States write to Dept CS, Penguin Books, 625 Madison Avenue, New York, New York 10022.

In Canada: For a complete list of books available from Penguins in Canada write to Penguin Books Canada Ltd, 41 Steelcase Road West, Markham, Ontario.

A Pelican Book

CONCEPTS OF MODERN MATHEMATICS

IAN STEWART

Most parents and many teachers are confused by 'modern maths', with its array of novel terms and symbols. To the layman, groups, sets and subsets mean little, and topology and Boolean algebra exactly nothing.

To dissipate some of the fog, Ian Stewart has written, not a handbook of modern mathematics, but a unique 'teach-in' to explain its aims, methods, problems and applications. His book sets out to define the principal concepts of modern mathematicians – their ways of looking at figures, functions and formulas. Since he is largely concerned with pure mathematics, his text inevitably demands some concentration, though only a smattering of algebra, geometry and trigonometry.

'Modern maths' looks askance at rules of thumb and Euclid (with his eternal triangle) conned by rote. With plenty of humour and a number of anecdotes Dr Stewart shows here how the new approach, once grasped, encourages a genuine understanding of mathematics.

A Pelican Book

ATOMS AND THE UNIVERSE

G. O. JONES, I. ROTBLAT, G. J. WHITROW

Third Revised Edition

Everything observed by the physicist or astronomer – on earth or in the sky – is packed with atomic matter and atomic energy. During the last two decades our knowledge about the universe has progressed by leaps and bounds, and there are signs that reasonably consistent theories about the universe are emerging.

In *Atoms and the Universe* two physicists and an astronomer survey the whole, vast field of physical science and modern astronomy, giving a complete and up-to-date guide to the structure of matter and the age and origins of the universe. They cover, among many subjects, the new accelerators developed by atomic physicists; the possibility of controlled thermonuclear reactors; the latest findings from artificial satellites and rockets; the speculations caused by quasi-stellar radio sources and exploding nebulae, and the important new work on the origin of elements and the evolution of stars.

This edition has been extensively revised and rewritten. Considered by many to be the outstanding work in its field, *Atoms and the Universe* will appeal not only to scientists, but to the layman with an intelligent interest in the nature of things.